The HLA System

Monographs in Human Genetics

Vol. 7

Editors: *L. Beckman*, Umeå and *M. Hauge*, Odense

S. Karger · Basel · München · Paris · London · New York · Sydney

The HLA System

An Introductory Survey

A. Svejgaard, M. Hauge, C. Jersild, P. Platz,
L.P. Ryder, L. Staub Nielsen and M. Thomsen

Tissue-Typing Laboratory of the Blood Bank and Blood Grouping Department, State University Hospital of Copenhagen, Copenhagen, and Institute of Clinical Genetics, University of Odense, Odense

14 figures and 17 tables, 1979

Second, revised edition

S. Karger · Basel · München · Paris · London · New York · Sydney

Monographs in Human Genetics

National Library of Medicine Cataloging in Publication
    The HLA system: an introductory survey
    A. Svejgaard ... [et al.] – 2nd rev. ed. – Basel, New York, Karger, 1979.
    (Monographs in human genetics; v. 7)
    1. HL-A Antigens   I. Svejgaard, Arne   II. Series
    W1 MO567P v.7 1979/WO 680 Hlll
    ISBN 3–8055–3049–8

© Copyright 1979 by S. Karger AG, 4011 Basel (Switzerland), Arnold-Böcklin-Strasse 25
Printed in Switzerland by Thür AG Offsetdruck, Pratteln
ISBN 3–8055–3049–8

Contents

Acknowledgements

The original work reported in this survey was aided by grants from the Danish Medical Research Council, the Danish Blood Donor Foundation, the Danish Rheumatoid Foundation, and the Foundation of Sept. 28th 1972. We are indebted to Dr. *Niels Holm,* Odense, and Dr. *Niels Morling* for helpful and critical comments during preparation of the manuscript, and to *Elly Andersen* and *Elisabeth Schacht* for excellent secretarial assistance.

Our thanks are also due to Professor *Erik Freiesleben* and Docent *Erna Möller* for invaluable discussions.

We owe a debt of gratitude to Dr. *Robin G. Harvey,* British Museum (Natural History), and Dr. *Coleman Smith,* University of Queensland for very valuable help during the production of the revised edition.

Table I. HLA Synonyms

HLA-A = LA = 1st series		HLA-B = FOUR = 2nd series		HLA-C = AJ = 3rd series	
new	previous	new	previous	new	previous
HLA-A1	HL-A1	HLA-B5	HL-A5	HLA-Cw1	T1,AJ
HLA-A2	HL-A2	HLA-B7	HL-A7	HLA-Cw2	T2,170
HLA-A3	HL-A3	HLA-B8	HL-A8	HLA-Cw3	T3,UPS
HLA-A9	HL-A9	HLA-B12	HL-A12	HLA-Cw4	T4,315
HLA-A10	HL-A10	HLA-B13	HL-A13	HLA-Cw5	T5
HLA-A11	HL-A11	HLA-B14	w14	HLA-Cw6	T7,IIP
HLA-A25	w25 ⎤ A10	HLA-B15	w15		
HLA-A26	w26 ⎦	HLA-B17	w17		
HLA-A28	w28	HLA-B18	w18	HLA-D = MLC series	
HLA-A29	w29	HLA-B27	w27		
HLA-Aw19	w19	HLA-B37	TY	new	previous
HLA-Aw23	w23 ⎤ A9	HLA-B40	w10		
HLA-Aw24	w24 ⎦	HLA-Bw4	w4,4a	HLA-Dw1	LD-101,W5a,J,PF
HLA-Aw30	w30	HLA-Bw6	w6,4b	HLA-Dw2	LD-102,7a,Pi,SKY
HLA-Aw31	w31	HLA-Bw16	w16	HLA-Dw3	LD-103,8a,SR
HLA-Aw32	w32	HLA-Bw21	w21	HLA-Dw4	LD-104,w15a,L
HLA-Aw33	w19.6	HLA-Bw22	w22	HLA-Dw5	LD-105,SA.1,IV
HLA-Aw34	Malay 2	HLA-Bw35	w5	HLA-Dw6	LD-106,Pr,VIII,pm.
HLA-Aw36	Mo*	HLA-Bw38	w16.1 ⎤ Bw16	HLA-Dw7	LD-107,12a
HLA-Aw43	BK	HLA-Bw39	w16.2 ⎦	HLA-Dw8	LD108,w10a
		HLA-Bw41	Sabell,MK	HLA-Dw9	TB9,OH
		HLA-Bw42	MWA	HLA-Dw10	LD16
		HLA-Bw44	B12 (not TT*) ⎤ B12	HLA-Dw11	LD17
		HLA-Bw45	TT*,4A2* ⎦		
		HLA-Bw46	HS,Sin2		
		HLA-Bw47	407*	HLA-DR series	
		HLA-Bw48	KSO,JA		
		HLA-Bw49	Bw21.1,SL-ET ⎤ Bw21	new	previous
		HLA-Bw50	Bw21.2,ET* ⎦		
		HLA-Bw51	B5.1	HLA-DRw1	
		HLA-Bw52	B5.2	HLA-DRw2	
		HLA-Bw53	HR	HLA-DRw3	
		HLA-Bw54	Bw22J	HLA-DRw4	~ HLA-D numbers
				HLA-DRw5	
				HLA-DRw6	
				HLA-DRw7	

Adapted from WHO (337). New = latest designation, previous = earlier designation; HLA = human leukocyte A (= first); ABC series = HLA-A + B + C, i.e. the classic antigens present on most cells and detected by serological methods; D = MLC = mixed leukocyte culture antigens; DR = HLA-D related antigens detected by serological methods, correspond to the Ia (immune region associated) antigens in mice.

Note on Nomenclature

We should perhaps apologize for starting this survey with some comments on the nomenclature. However, as this has been one of the major problems for non-HLA workers trying to understand HLA, and as a rather pronounced change of the nomenclature has been made very recently, we find it necessary to give this note at a place where it can be easily consulted.

The various HLA antigens originally received preliminary local designations when they were discovered, but since 1967, a WHO Nomenclature Committee has steadily tried to ascribe uniform designations to the antigens when it was felt that they were adequately defined. At first, the antigens were named 'HL-A' followed by a number (e.g. HL-A1) without taking into account to which segregant series the antigens belonged (335). Very recently, however, a marked change was introduced (336) by calling the segregant series A, B, C, and D, and by moving the hyphen from 'HL-A' to 'HLA-'. In the latest report from the Nomenclature Committee (337), the so-called B lymphocyte alloantigens were named HLA-DR (for HLA-D related) antigens because of their close relationship with the HLA-D antigens. Table I gives the most recent nomenclature (337) and lists some synonyms for the various antigens. It has not been possible to include all the designations used originally by the individual investigators, but these can be found in the joint reports of the previous workshops and in earlier surveys (12, 132, 133, 149).

The antigens of the A and B series have retained their original numbers (except W5 and W10 which overlapped with other antigens), and it was decided that there should be no overlapping in numbers between these two series. However, the C and D series both have numbers from 1 upwards, so it is necessary to give the letters of the series in pheno- and genotypes: e.g., HLA-A1,2;B5,8;Cw1 and *HLA-A1,B8,C-/A2,B5,Cw1*, respectively.

The letter 'w' (for workshop) indicates antigens which are not felt to be completely well-defined. The w-antigens are going to maintain the same number when they are eventually fully accepted. Accordingly, it does not seem absolutely logical to use w designations, and for simplicity we are going to omit the w's in many places in this survey. Note, however, that if this is done for the C series, confusion with complement factors may arise.

1. Introduction

The polymorphism of the HLA system is unique in man both in terms of genetics and as regards biological functions. The extreme genetic polymorphism of the system is exemplified by the fact that it can in theory give rise to many million different HLA phenotypes at least in a Caucasian population. Moreover, each of these phenotypes are individually rare, the most common will in many parts of the world have a frequency of less than 0.5%. In comparison, the most frequent phenotype in another very polymorphic genetic system in man, the Rhesus blood groups, has a frequency as high as 30% in many places.

In terms of biological functions, the HLA system is also outstanding because it controls so many diverse characters. Being first recognized as a leukocyte blood group system, it was soon realized that it is the major histocompatibility system in man: only if HLA compatibility is present will a transplanted kidney have a high chance (90% or more) of survival. However, it has more recently been found that the biological significance of HLA goes far beyond that of transplantation. Firstly, HLA and related systems in animals play a fundamental role in the cooperation between various cells of the immune system and thereby control the immune response to a variety of antigens, and secondly, it also controls some of the components of the complement cascade. Thus, studies of these systems have greatly expanded our knowledge on the function of the immune system. Moreover, it appears that some diseases occur preferentially in individuals possessing certain HLA factors. These associations between HLA and disease are in general much stronger and thus more biologically significant than those which have previously been found between blood groups and disease. For example, in ankylosing spondylitis – an arthropathy of the spine – the association with an HLA factor is so strong that HLA typing can be used diagnostically. For several other diseases, e.g., multiple sclerosis and juvenile diabetes mellitus, the observation that they are associated with HLA have created promising new leads in the studies of pathogenesis and etiology, which in turn may permit better therapeutic and prophylactic measures than those presently applied.

From a genetic point of view, the HLA system offers a new and very effective tool in many studies of fundamental importance, such as inheritance of disease susceptibility, anthropology, natural selection and the interaction between closely linked genes. The maintenance of the enormous HLA polymorphism is an exciting mystery and a challenge to many geneticists, but the solution to this and many other unsolved problems related to the biological functions of this system is likely to come by combining the results of a diversity of studies involving a multitude of scientific areas such as immunology, biochemistry, genetics, and medicine.

Accordingly, the HLA system is of immediate interest and relevance to many workers in medicine and most other fields related to human biology, and it is our hope that this short survey may give a useful outline of our present knowledge of HLA for those who are not themselves directly involved in HLA work. We are well aware that it has been written during a very active phase of research within this field and that the next few years are likely to yield important new information and change many of the views and ideas expressed with more or less reservation here. However, we still find it worthwhile to summarize what is known at the present time as it seems clear that few fields of human biology will remain unaffected by the achievements in the HLA area.

The list of references is not in any way intended to cover the entire literature (which is already very abundant) but for each major topic we have tried to select a few key reviews or reports which should be consulted in case more detailed information is required.

2. Historical Background

The first evidence in favor of the existence of human leukocyte blood groups was put forward in 1954 by *Dausset* (61) who observed that patients whose sera contained leukoagglutinins had received a larger number of blood transfusions than patients lacking such antibodies. This observation indicated that these antibodies were not autoantibodies as was previously thought, but rather isoantibodies (alloantibodies) induced by the infusion of cells carrying isoantigens not present in the recipient. In 1958, *Dausset* (62) provided strong support for this assumption when he observed that sera from seven polytrans-fused patients agglutinated leukocytes from about 60% of the French population but not the leukocytes of the seven patients. *Dausset* termed the leukocyte isoantigen MAC (now HLA-A2 + A28) and this was the first discovery of an HLA antigen. *Dausset and Brecy* (63) also provided evidence from twin studies that leukocyte isoantigens are genetically determined, and family studies by *Payne and Rolfs* (221) substantiated this interpretation. These authors also showed that pregnancies may induce the formation of leukocyte isoantibodies as did *van Rood et al.* (245) independently. *Van Rood* (241) fought the imperfect serology of this early stage by computer analysis and discovered the diallelic leukocyte antigen system, 4a, 4b (now Bw4 and Bw6).

By means of absorption studies, *van Rood* (241) also found that these antigens are present on most human tissues. Simultaneously, *Shulman et al.* (268, 269) showed that typing for leukocyte antigens can also be done by means of a reliable complement-fixation test with blood platelets as antigen source. However, sera reacting sufficiently strongly in this test are unfortunately rare, and it was therefore of great importance, when *Terasaki and McClelland* (303) in 1964 introduced the more sensitive microlymphocytotoxicity test which in various modifications has maintained its major role in serological HLA typing until the present time.

Following the discovery of the first leukocyte antigens the number increased rapidly, and in 1965, *Dausset et al.* (70) and *van Rood et al.* (247) suggested on the basis of population studies that most of these antigens belong to one and the same genetic system. Family studies supporting this assumption were published in 1966 by *van Rood et al.* (244) and *Dausset et al.* (69) and the proof was obtained during the 3rd Histocompatibility Workshop in 1967, when

16 different teams typed the same sample of Italian families selected by *Ceppellini et al.* (50, 52). The close genetic relationship of the determinants of these leukocyte antigens was so convincing that the term HLA [the first *(A)* *H*uman *L*eukocyte antigen system discovered] was approved by a WHO nomenclature committee (335) for this genetic system.

The exact arrangement of the genes within the corresponding chromosomal region was still an open question although evidence of two mutational sites, each with at least three alleles had been provided by *Ceppellini et al.* (52). With the discovery of the disturbing fact that many if not most HLA antibodies are in fact cross-reacting (64, 293), i.e. the same antibody reacts with two or more different antigens, it became possible for *Kissmeyer-Nielsen et al.* (146) in 1968 to present a simplified model of the HLA system as being composed of two closely linked loci with genes controlling two segregant series (LA and FOUR, now A and B) of antigens. This model was substantiated by *Staub Nielsen*'s (147) discovery of a family showing crossing-over between the two loci and further proof was obtained by classical genetic analyses in joint Danish-Norwegian population and family studies (287, 298). The 4th Histocompatibility Workshop (132) brought the final confirmation of the two segregant series. The existence of a third segregant series (AJ, now C) strongly associated with the B series, was already then suggested by *Sandberg et al.* (258) but was not taken seriously until other antigens fitting into this series were discovered (294). The first case of crossing-over between the C and the B series was found by *Löw, et al.* (177) who at the same time established that C is between the A and B loci. *Bernoco et al.* (31) showed by means of an elegant method developed by *Kourilsky et al.* (156) that A and B series antigens exist as separate molecular entities on the cell surface membrane, and *Solheim et al.* (278) showed that C antigens also exist as separate molecules (cf. p. 40).

The basis for the extreme interest in the HLA system obviously was the concept that this is the major histocompatibility system in man. Such systems have been known in other vertebrates since 1948 (152) and the proof that HLA is the strongest histocompatibility system of man was obtained in 1968 by *Ceppellini et al.* (53) and *Amos et al.* (11) in skin graft experiments: skin transplanted between HLA-identical siblings survives much longer than skin from HLA-dissimilar siblings. More important, it was also found that kidneys from HLA-identical sibling donors almost always have a perfect survival in the recipient, and there were thus great expectations concerning the possibility of improving the survival of kidneys from unrelated donors by HLA matching (242). However, these expectations have only been met with moderate success (cf. p. 59).

Another line of research aiming at improving graft survival arose from the discovery in 1964 of *Bain and Lowenstein* (18) and *Bach and Hirschhorn* (16) that lymphocytes from two different individuals undergo blast transformation and divide when mixed and cultured *in vitro* (the so-called mixed lymphocyte culture test, MLC; cf. p. 14). In 1967; *Bach and Amos* (15) showed that the MLC test gave negative reactions when leukocytes from HLA-identical siblings were mixed together. This gave strong support to the assumption that the reaction is controlled by genes closely linked to or identical with those controlling the 'classical' HLA antigens of the A, B, and C series. *Sørensen et al.* (148, 300) performed MLC tests between unrelated individuals who were identical for these antigens and found stimulation indicating that the known A, B, and C antigens could not be solely responsible for the reactions. *Yunis et al.* (344) made observations suggesting that a separate locus exists which controls MLC and which is separate from A and B, closer to the latter and not located between these two loci. The close proximity of the MLC (HLA-D) locus with the B locus was substantiated by *Dupont et al.* (85). Finally, in 1973, it proved possible to type for HLA-D determinants by means of HLA-D homozygous cells (84, 137, 189, 328), and the polymorphism at the HLA-D locus started to become unravelled. The significance of HLA-D typing and cross-matching in transplantation with unrelated kidney donors seems to be greater than that obtained by matching for HLA-ABC antigens (310), and HLA-D matching seems to be of major importance in bone marrow grafting (81, 101a, 171).

A very important step forward was made in 1973 by *van Leeuwen et al.* (169) who provided evidence that it is possible to type serologically for some antigens very closely related to the HLA-D antigens. The same investigators (246) developed a suitable method for the detection of these antigens which are now called HLA-DR (for HLA-D related) antigens, and it now seems as if compatibility for HLA-DR antigens is associated with a successful outcome of cadaver kidney transplantation (7, 179, 223, 319).

A completely different field of research gradually revealed that some diseases occur preferentially in individuals carrying certain HLA antigens, and this has opened new aspects in diagnosis and in the study of genetics and etiology and pathogenesis in a variety of disorders.

The stimulus came from the already existing knowledge about the importance of the HLA analogue, the H-2 system, for the development of viral leukemia in mice described by *Lilly et al.* (175) in 1964. HLA had barely been recognized as a system before the first studies on HLA antigen frequencies in human lymphoid malignancies were reported by *Amiel* (9) and by *Kourilsky et al.* (155). One of them showed no association between HLA and acute lymphatic leukemia (155) while the other (9) gave evidence of an association between HLA and Hodgkin's disease and provoked a number of other investigations of such patients. In 1972, it became clear that there is no strong association for that disease (198), but at the same time strong associations were found for psoriasis (251) and celiac disease (93), and the following year the strongest association observed until now, between B27 and ankylosing spondylitis, was discovered

(41, 262). A number of other associations has been found, and it has been documented that a disease may be more strongly associated with MLC determinants than with the classical ABC determinants (130). The existence of these associations shows that the HLA system plays a much broader role in biology than that related to transplantation.

Several of the associations observed may be seen in relation to the discovery (185) that the H-2 system of mice contains so-called immune response (Ir) genes which control the ability of the individual animal to respond with specific antibody formation upon challenge with certain antigens (183). This has lead to a search for similar genes within the HLA complex. The observation of *Levine et al.* (172) that ragweed hay fever seems to be inherited in individual families together with certain HLA complexes was the first piece of evidence favoring this assumption. Another hint has been obtained by *Thomsen et al.* (312) who found that antiadrenal antibodies were seen significantly more often in HLA-B8-positive than in B8-negative patients with idiopathic Addison's disease. More recently, evidence has been provided that HLA-D antigens are intimately involved in the immune response because they determine the cooperation between monocytes and T lymphocytes (30, 112). Moreover, the HLA-ABC antigens seem also to play a fundamental role in the lysis of virus-infected and hapten-conjugated cells by T killer lymphocytes (79, 103, 188). These findings confirm earlier exciting observations made by *Doherty and Zinkernagel* (80), *Zinkernagel and Doherty* (347) and by *Shearer* (264) in mice, and are key points to our understanding of the biological significance of the HLA system.

The discovery that certain components of the complement cascade (8, 99, 237, 302) are controlled by HLA adds a new dimension to the biological function of this system.

Thus, even if many of these diverse fields of research related to HLA have just been opened quite recently and although some of the results may need confirmation, it is beyond doubt that the implications of the continuously increasing knowledge about the details of the HLA system will improve our understanding of human biology including medicine very greatly. As evidenced by this brief account of the history of HLA, a period of only about 15 years has brought us a tremendous amount of new information on very basic properties of some fundamental components of the human organism.

3. Components of the HLA System

It is self-evident but not always fully appreciated that the methods available for study set a limit for our recognition of nature's secrets. For example, 40 years went by after the discovery of the ABO system and thousands of newborns died or were crippled from hemolytic disease before the Rhesus system was found. The Rhesus antibodies responsible were there all the time but escaped detection as they only very rarely agglutinate Rhesus-positive erythrocytes suspended in saline which was used routinely. Only after the discovery of the Rhesus system was it realized that these antibodies need special conditions, e.g. the presence of colloid media, before they agglutinate the relevant erythrocytes. By analogy, our present concept of the HLA system obviously depends on the methods which have been in use until now.

The HLA system is as mentioned the most complex genetic system known to exist in man. It consists of many closely linked loci which control a variety of characters. The first characters which brought the system to light were alloantigens on human leukocytes, and the term 'HLA' is derived from the first *(A) Human Leukocyte* antigen system discovered (335). It is now clear that HLA is the major histocompatibility system (MHS) or complex (MHC) in man (314). There has been some discrepant use of the nomenclature: some investigators reserved the term 'HLA' for the 'classical' HLA antigens of the A, B, and C series, but we feel that HLA should cover all factors of related function coded by genes within close distance of the ABC genes, as most of these factors seem to have related functions. In fact neither of the terms MHC or HLA describes these functions fully.

The HLA system controls at least three different groups of characters: antigens, various immune responses, and some components of the complement cascade.

At the present time, most antigens of the HLA system are conveniently grouped into three categories: (1) one containing the *HLA-ABC antigens* present on most cells and readily detectable by currently used serological methods (the 'classical' or serologically detectable (SD) antigens); (2) one comprising the *HLA-D antigens* detected in the mixed lymphocyte culture (MLC) test (MLC antigens = lymphocyte-defined (LD) antigens); and (3) one including the *HLA-DR antigens* which are closely related — perhaps identical — to the HLA-D

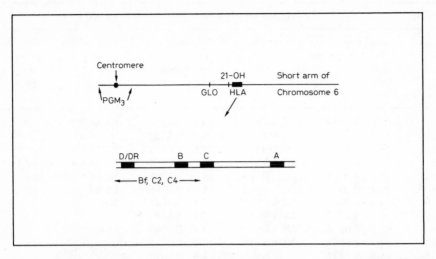

Fig. 1. Linkage relationship and arrangements of HLA genes on chromosome No. 6. The locus for phosphoglucomutase 3 (PGM$_3$) is about 20–25, and the GLO (glyoxylase) locus about 5 centimorgans from HLA. The locus for 21-hydroxylase (21-OH) is between the GLO and HLA-A loci. The loci for some components of complement (Bf, C2, and C4) are close to the D/DR and B loci, perhaps between them.

antigens, but which can be recognized by special serological methods. The HLA-D and HLA-DR antigens are mainly present on macrophages and B lymphocytes and correspond to the so-called Ia (immune region associated) antigens of mice.

The genetic loci carrying the HLA genes are located with a distance of a few centimorgans from one another on the short arm of chromosome No. 6 (98, 162, 280). The arrangement of HLA loci on this chromosome is shown in figure 1. The A, B, and C loci control classical HLA antigens and the D locus controls the D antigens. The exact position of the DR locus (or loci) is unknown, but it must be very close to (perhaps identical with) D. The Ir loci and those controlling the complement components C2, C4, and Bf are probably also close to the D locus.

It has been suggested that, judged from the recombination frequency of about 1% between the A and B loci, there may be several hundreds or even thousands of genes between these two loci (37).

Tables II and VII give the frequencies of various HLA characters and the corresponding genes in Danes who are quite representative of North Europeans. All of the antigens are inherited as simple codominant characters which are present from early fetal life (51) and do not change with age. There may still be

Table II. Frequencies of the antigens and genes of the HLA-A, B and C series in Danes

HLA-A series			HLA-B series			HLA-C series		
antigen	frequency		antigen	frequency		antigen	frequency	
	anti-gen %	gene		anti-gen %	gene		anti-gen %	gene
HLA-A1	32.0	0.1751	HLA-B5	10.1	0.0521	HLA-Cw1	5.7	0.0289
HLA-A2	53.3	0.3163	HLA-B7	25.8	0.1388	HLA-Cw2	9.7	0.0496
HLA-A3	28.2	0.1529	HLA-B8	24.6	0.1317	HLA-Cw3	35.0	0.1938
HLA-A9	17.3	0.0906	HLA-B12	25.7	0.1382	HLA-Cw4	17.1	0.0897
Aw23	3.3	0.0166	Bw44	24.4	0.1306	HLA-Cw5	n.i.	
Aw24	16.4	0.0856	Bw45	1.5	0.0076	HLA-Cw6	33.1	0.1820
HLA-A10	9.8	0.0505	HLA-B13	4.4	0.0221	blank		0.4560
Aw25	4.6	0.0231	HLA-B14	3.8	0.0193			
Aw26	5.2	0.0266	HLA-B15	18.4	0.0965	total		1.000
HLA-A11	10.9	0.0559	HLA-Bw16	5.9	0.0298			
HLA-A28	8.5	0.0437	Bw38	1.8	0.0090			
HLA-Aw19	20.3	0.1071	Bw39	3.8	0.0191			
Aw29	4.5	0.0266	HLA-B17	8.0	0.0409			
Aw30	n.i.		HLA-B18	7.7	0.0391			
Aw31	n.i.		HLA-Bw21	3.6	0.0184			
Aw32	6.1	0.0312	HLA-Bw22	3.9	0.0197			
Aw33	n.i.		HLA-B27	9.4	0.0483			
Blank		0.0079	HLA-Bw35	14.6	0.0759			
			HLA-B37	2.6	0.0129			
Total		1.0000	HLA-B40	18.4	0.0969			
			HLA-Bw41	2.0	0.0103			
			HLA-Bw47	0.4	0.0021			
			blank		0.0070			
			total		1.0000			

The data are mostly from *Staub Nielsen et al.* (210, 211) and were obtained from investigations of abou 400–3,000 unrelated Danes. Frequencies of antigens are given as percentages, while those for the correspond ing genes are simple frequencies. The *HLA-Aw25* and *Aw26* genes comprise together the *HLA-A10* gene (i.e an *HLA-A10* gene is either *Aw25* or *Aw26*). A similar relationship exists for *Aw23* and *Aw24* whic comprise *A9*, and for *Bw38* and *Bw39* which comprise *Bw16*. In analogy, *Aw19* is made up of *A29*, *Aw30 Aw31*, *Aw32* and *Aw33*. Accordingly, the frequencies of the following genes are not included in the sums *Aw23*, *Aw24*, *Aw25*, *Aw26*, *A29*, and *Aw32* of the A series, and *B12*, *Bw38*, and *Bw39* of the B series. n.i. Not investigated.

Antigen frequencies are calculated by the number of individuals having the antigen divided by the tota number of individuals investigated. As there are many heterozygotes (e.g. of the genotype *HLA-A1/A2*), th sum of the antigen frequencies from a given series is more than 100% (but can never exceed 200%) becaus heterozygotes are counted twice (e.g. both for HLA-A1 and A2). *Gene frequencies* indicate the number o genes of a certain type divided by the total number of alleles studied, i.e. twice the number of individuals; a there can be only one gene from a series on one chromosome, the sum of gene frequencies must for one serie (locus) be one. The *'blank' (null)* genes indicate genes the products of which cannot yet be recognized.

'blank' or 'null' genes at the A and B loci, and there are such genes at the C, D, and DR loci as we cannot (yet) recognize all gene products.

For each of the loci there are $(^1/_2)n(n + 1)$ possible genotypes (n being the number of alleles) corresponding to $(^1/_2)(n^2-n + 2)$ different phenotypes, and when the 15, 21, 7 and 9 different alleles at the A, B, C, and D/DR loci, respectively, are combined, there is a theoretical probability of finding about 20 million different phenotypes in the population. Only a minority of these have actually been observed and some may not exist at a given time. However, it is a fact that all HLA phenotypes are individually rare and that it is accordingly difficult to find two unrelated individuals who have identical HLA phenotypes; no other genetic system in man shows a similarly high degree of polymorphism.

3.1. A, B, and C Segregant Series (ABC Series, the 'Classical' or 'SD' HLA Antigens)

These antigens are usually detected on peripheral blood lymphocytes by means of alloantibodies present in sera from individuals immunized by pregnancies or transfusions. The so-called lymphocytotoxic test generally used for the typing for these antigens is described in section 11.1.2.

The classical HLA antigens fall into three *segregant series* because they within each series are inherited as if controlled by allelic genes. The three series are called the A (LA or FIRST), B (FOUR or SECOND), and C (AJ or THIRD) segregant series following their order of discovery. The old designations LA, FOUR and AJ have historical backgrounds: LA refers to the LA1, LA2 system discovered by *Payne et al.* (220), FOUR to *van Rood's* (241) 4a,4b antigens, and AJ are the initials of the Swedish woman whose serum first allowed detection of a C series antigen (258).

The concept of three segregant series is based on the following evidence obtained from population and family studies (5, 52, 65, 180, 182, 298, 317).

(1) No individual possesses more than two antigens from each of the three segregant series. Whenever individuals with three antigens from the same series (often called 'triplets') have been reported, technical typing difficulties have been the most likely explanation.

(2) Perhaps the most convincing evidence in favor of the existence of the three separate segregant series is the observation of recombination between them. Table IV shows a family which demonstrates a crossing-over between the A series on one hand and the B and C series on the other and several such families have been found. Furthermore, two families have recently been observed (114, 177) in each of which a recombination had occurred between the A and C series on one side and the B on the other, and thus the sequence of the series on the chromosome must be: A:C:B or vice versa.

Table III. Inheritance of HLA antigens within a family

| | Antigens | | | | | | | | Phenotype | Genotype (haplotype/haplotype) | |
| | A series | | | | B series | | | | | full | abbreviated |
	HLA-A1	A2	A3	A9	B5	B7	B8	B12			
Father	+	–	+	–	–	+	+	–	HLA-A1,3;B7,8	*HLA-A1,B8/A3,B7*	a/b
Mother	–	+	–	+	+	–	–	+	HLA-A2,9;B5,12	*HLA-A2,B5/A9,B12*	c/d
1st child	+	+	–	–	+	–	+	–	HLA-A1,2;B5,8	*HLA-A1,B8/A2,B5*	a/c
2nd child	+	–	–	+	–	–	+	+	HLA-A1,9;B8,12	*HLA-A1,B8/A9,B12*	a/d
3rd child	–	+	+	–	+	+	–	–	HLA-A2,3;B5,7	*HLA-A3,B7/A2,B5*	b/c
4th child	–	+	+	–	+	+	–	–	HLA-A2,3;B5,7	*HLA-A3,B7/A2,B5*	b/c

Plus and minus indicate presence and absence, respectively, of the antigen in question. Note that in the children, the paternal haplotype is usually given before the maternal. In the phenotypes, A series antigens are given before B series antigens. No agreement has yet been made as to where C series antigens should be put; we usually give them at the end although it would seem more logical to put them between the other two.

Table IV. Crossing-over within the HLA system

| Relative | HLA antigens | | | | | | | | | Phenotype | | | Genotype | |
| | A series | | | B series | | | | C series | | | | | A,C,B/A,C,B | |
	A1	A2	A3	B7	B8	B22	B27	C1	C2	A	B	C		
Father	+	+	–	+	–	–	+	–	+	1,2	7,27	2	*1,–,7/2,2,27*	a/b
Mother	+	–	+	–	+	+	–	+	–	1,3	8,22	1	*1,–,8/3,1,22*	c/d
1st child	–	+	+	–	–	+	+	+	+	2,3	22,27	1,2	*2,2,27/3,1,22*	b/d
2nd child	+	–	+	+	–	+	–	+	–	1,3	7,22	1	*1,–,7/3,1,22*	a/d
3rd child	+	+	–	–	+	–	+	–	+	1,2	8,27	2	*2,2,27/1,–,8*	b/c
5th child	+	–	+	–	–	+	+	+	+	1,3	22,27	1,2	*1,2,27/3,1,22*	a:b/d
6th child	+	–	–	+	+	–	–	–	–	1	7,8	–	*1,–,7/1,–,8*	a/c

Plus and minus indicate that an antigen is present and absent, respectively. Only the antigens present in the family have been shown. Letters A, B, and C have been omitted from the phenotypes and genotypes for simplicity. Small letters *a, b, c,* and *d* indicate the parental HLA haplotypes. This family is the CPH-41 family (289) and has also been used for MLC studies (85).

(3) When two antigens from the same series are present together in a parent, one of them – never more or less – is handed on to the child. On an average, half of the children inherits one antigen and the remaining the other antigen, i.e. the *segregation ratio* is 0.5, and when larger family materials have been studied, segregation ratios of 0.5 have been found for all antigens (5, 132, 180, 291, 294, 298, 317).

(4) If factors are controlled by allelic genes, a random mating population should show *Hardy-Weinberg equilibrium* (49, 174) for these factors. This has been shown to be true for each of the three segregant series in quite large series of unrelated individuals (135, 210, 298).

(5) Further reasons to regard the three series as separate entities arise from the observations that serological *cross-reactions* (cf. p. 84) between various antigens occur almost exclusively within a given series – A, B, or C – not between them. This point was first noted in 1969 (287) and since then only two possible exceptions to this rule have been found (27, 170).

(6) As discussed on page 40, the antigens from each series have been shown by 'capping' and 'stripping' experiments to be present on separate molecules on the cell surface (31, 182, 278).

It appears from table II that there may still be some unknown antigens of the A and B series, even in Caucasians who have been studied most extensively, and in the C series the frequency of the *'blank'* or *'null'* alleles is still high. When more HLA antibodies are found, these antigens are likely to be defined.

Due to the close linkage, the A, B, and C characters will be transmitted 'en bloc'. Table III shows the inheritance of HLA antigens in a family. For simplicity only the A and B series antigens are shown. It appears that the paternal antigens A1 and B8 travel together to the first two children, while the last two sibs have received A3 and B7 from the father. Accordingly, the *A1* and *B8* genes are present on the same chromosome, or in the same haplotype (52), whereas *A3* and *B7* constitute the other paternal HLA haplotype. By analogy, the maternal HLA haplotypes are *A2,B5* and *A9,B12*. Frequently, when analyzing such data, small letters (*a, b, c* and *d*) are used to indicate the parental haplotypes. The term *haplotype* was introduced by *Ceppellini et al.* (52) to indicate the genes found on one and the same of the two homologous chromosomes. In this example, siblings number 3 and 4 have inherited the same parental haplotypes (*b* and *c*), and thus they are *HLA identical*. As recombinations are rare within the HLA system (cf. below), such HLA-identical siblings are in most cases identical not only for the HLA factors known by now, but also for those which cannot yet be detected. Siblings 1 and 2 differ by one haplotype (*c* and *d*) and are called *haploidentical*. In contrast, the second child has no haplotypes in common with the third (and fourth), and this pair of siblings is as different as random unrelated individuals in respect of HLA.

Table IV shows the HLA types of a family in which a recombination has

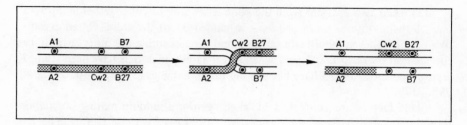

Fig. 2. Graphical illustration of the crossing-over as it took place during meiosis in the father in table IV. The recombinant *HLA-A1,Cw2,B27* haplotype was inherited by the fifth child.

occurred between the A series on one hand and the B and C series on the other. Figure 2 illustrates in a simplified way how this crossover took place during meiosis in the father.

3.2. HLA-D = Mixed Lymphocyte Culture (MLC) Determinants

When lymphocytes from two individuals are mixed and cultured *in vitro*, they will usually stimulate each other to divide (16, 18, 299) which can be measured as described on page 15. This MLC reaction is due to the fact that lymphocytes have both 'antigens' and receptors for foreign 'antigens' on their surface. The 'antigens' responsible for the stimulation are the HLA-D or MLC determinants, previously called LD (lymphocyte defined) or LAD (lymphocyte activating determinant) antigens. In the one-way MLC (17, 139) test (fig. 3), the lymphocytes from one of the two individuals have been prevented from dividing, e.g. by X-irradiation.

Although the HLA-D determinants or 'antigens' have not (yet) been formally shown to be antigens (in the sense of being able to give rise to circulating antibodies), we use for simplicity mostly the term HLA-D 'antigen' instead of the perhaps more strictly correct HLA-D 'determinant'. Moreover, the results obtained with the so-called primed lymphocyte typing (PLT) test (265) discussed on pages 18 and 90 indicate that HLA-D determinants are indeed antigens.

As mentioned earlier, the factors responsible for the stimulation in the MLC test are coded for by genes within the HLA gene complex. This must be so because lymphocytes from HLA-identical sibling pairs almost never stimulate each other in MLC — see *Sørensen* (299) for review — whereas all other sib combinations almost invariably show stimulation. Moreover, siblings differing by two haplotypes (say *a/c versus b/d*) stimulate each other to the same extent as unrelated individuals, about twice as much as do haploidentical siblings (*a/c*

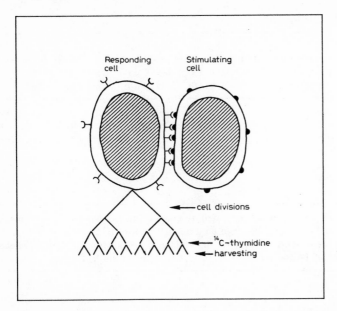

Fig. 3. The principle of the one-way MLC test. The stimulating cells have been prevented from dividing, e.g. by X-irradiation. The untreated responding cells recognize the foreign HLA-D antigens (black symbols) of the stimulating cell and are stimulated to divide. The degree of cell proliferation is measured by the uptake of ^{14}C-thymidine in the DNA of the cells. The responding cells are T lymphocytes (which lack HLA-D antigens), and the stimulating cells are B lymphocytes and monocytes.

versus *a/d* for example) and parent-child combinations which are of course always haploidentical.

Table V shows the results of MLC tests performed between all members of a family and an unrelated control and illustrates the absence of stimulation between HLA-identical siblings and the difference between a one- and a two-haplotype response.

It would have been expected that the antigens eliciting stimulation in the MLC test were those of the A, B, and C segregant series. On the contrary, however, there is quite solid evidence that this is not the case. Firstly, it appeared that unrelated individuals being identical for two A as well as two B series antigens (e.g. of the genotype HLA-A1,B8/A3,B7) mostly show MLC stimulation (148, 300) and this holds even when the identity is extended to include C series antigens (83). Secondly, it appeared from MLC studies in families with recombinants between the A and B locus that incompatibility for B series antigens is much more important than incompatibility for A series antigens (85, 90, 344). Thirdly, and most important, some cases of MLC stimulation

Table V. MLC test in a family

Responder		Stimulator						
relationship	HLA genotype	Fx *a/b*	Mx *c/d*	1x *a/c*	2x *a/c*	3x *b/c*	4x *b/d*	Ux *x/y*
Father	*a/b*	0	++	+	+	+	+	++
Mother	*c/d*	++	0	+	+	+	+	++
1st child	*a/c*	+	+	0	0	+	++	++
2nd child	*a/c*	+	+	0	0	+	++	++
3rd child	*b/c*	+	+	+	+	0	+	++
4th child	*b/d*	+	+	++	++	+	0	++
Unrelated	*x/y*	++	++	++	++	++	++	0

The test is a 'chess-board' between untreated responder and x-irradiated (x) stimulator lymphocytes from all family members and an unrelated control (U). Letters *a, b, c, d, x*, and *y* indicate different HLA haplotypes determined by ABC typing. Zero's indicate absence and one and two plusses various degrees of stimulation in the MLC tests.

between HLA-identical sibling pairs were discovered (90, 344). It might be argued that such stimulation could be due to incompatibility for MLC antigens controlled by other systems less polymorphic than HLA. One such system (the M system) not linked to H-2 has been found in mice (94). However, this system is not very polymorphic and it has been shown for some of the human cases that this is a very unlikely possibility: the findings can almost only be explained by assuming the existence of a separate MLC locus (HLA-D) close to the B locus and outside the interval between the A and B loci. Finally, the recent establishment of HLA-D typing has yielded results which show quite clearly that HLA-D antigens are different from those of the A, B, and C series (cf. below). An excellent review on MLC determinants in man has recently been published by *Dupont et al.* (82).

In addition to this so-called 'strong' MLC locus (D or LD-1 locus), the products of which cause strong stimulation there is some evidence in favor of the existence of much weaker MLC determinants controlled by genes at a locus (LD-2) located in the neighborhood of the A locus (32, 90, 316). However, because of their weakness — or the insensitivity of available methods — these determinants are so ill-defined at the present time that in the following we shall deal only with the strong ones if not otherwise stated. ABO blood group antigens do not cause detectable stimulation in MLC (299).

It has been found that the cells stimulating in MLC are primarily B lymphocytes and monocytes, while the responding cells belong to the class of T lymphocytes (111, 205, 299) (cf. p. 45).

Table VI. Principle in HLA-D typing

Responder	Stimulator genotype					Phenotype of responder
	D1/D1	D2/D2	D3/D3	D4/D4	D5/D6	
1	0	++	0	++	++	D1,3
2	++	0	++	0	++	D2,4
3	0	++	++	++	++	D1
4	++	++	++	++	++	?

Responders 1–4 are 'unknown' individuals to be typed. D1/D1, D2/D2, D3/D3, and D4/D4 are various HLA-D-homozygous x-irradiated typing cells; D5/D6 are heterozygous control cells. Plusses indicate stimulation and zero's absence of or (usually) low stimulation. The interpretation of the results in terms of responder phenotypes is shown to the right.

Table VII. Frequencies of HLA-D and HLA-DR antigens and genes in Danes

HLA-D series			HLA-DR series		
antigen	frequency		antigen	frequency	
	antigen %	gene		antigen %	gene
HLA-Dw1	19.9	0.105	HLA-DRw1	19.3	0.101
HLA-Dw2	25.8	0.139	HLA-DRw2	29.4	0.160
HLA-Dw3	26.3	0.142	HLA-DRw3	23.9	0.127
HLA-Dw4	19.4	0.102	HLA-DRw4	28.4	0.154
HLA-Dw5	5.8	0.029	HLA-DRw5	n.i.	
HLA-Dw6	16.0	0.084	HLA-DRw6	n.i.	
HLA-Dw7	18.0	0.094	HLA-DRw7	15.6	0.081
HLA-Dw8	7.5	0.038	HLA-DRw8	23.4	0.127
Blank		0.267	blank		0.250
Total		1.000	total		1.000

HLA-D antigens were studied in about 200–350 and DR antigens in 109 unrelated Danes.

The principle of *HLA-D typing* in the MLC test by means of HLA-D homozygous typing cells is illustrated in table VI. HLA-D-homozygous *(a/a)* cells carry only one type of HLA-D antigen (a) on their surface, and thus they will not stimulate responder cells carrying themselves the a antigen, whether these responder cells are also homozygous *(a/a)* or heterozygous *(a/x)*. In contrast, responding cells lacking (a) can recognize this antigen as foreign and will respond to the a/a cells. A number of HLA-D determinants have been quite well-defined by this method within the last 6 years (134, 138, 312). Presently, about ten different HLA-D antigens are known, and the frequencies of some of the best characterized are shown for the Danish population in table VII, which also gives the corresponding gene frequencies (134, 312).

As is the case for the A, B, and C series antigens, no individual has yet been reported who definitely carried more than two of these HLA-D antigens indicating that they may be controlled by allelic genes.

Hardy-Weinberg tests have not shown deviations from expectation (134, 138, 309). The number of individuals typed for more than two HLA-D determinants is still too small to supply sufficient evidence in favor of simple allelism. In the few family studies known to us, back-cross matings (of the type $D1/D2 \times Do/Do$) have not given reasons to reject the assumption that these strong MLC determinants are controlled by genes at one locus. It is worth noting that the above design of HLA-D typing may lead to false assignment of an HLA-D antigen to an individual if his ability to respond is in some way impaired, e.g. because he is immunodeficient or because he develops suppressor or killer cells (cf. p. 54) in the culture which has been described in a few cases (159, 187, 306, 311).

A new approach to HLA-D typing was recently introduced by *Sheehy et al.* (265) who developed the so-called primed lymphocyte typing (PLT) test described on page 90. In this test, lymphocytes are primed specifically *in vitro* to give a rapid response to certain HLA-D antigens. Accordingly, it is possible to establish the HLA-D type of an individual by investigating whether his lymphocytes will induce a rapid response when mixed with cells primed specifically to the various HLA-D antigens. In contrast to HLA-D typing by means of homozygous typing cells, the PLT test allows assignment of HLA-D antigens on the basis of a positive response. The results obtained with these two methods are not always identical (260, 311, *Morling* unpublished) which may be partly due to cross-reactivity of primed cells, and for the time being it may be better to use the term 'PLT antigens' for the determinants detected by the PLT test.

As often occurs in science, a theory can rarely be proven, it is rather maintained on the grounds of incomplete evidence against it. We feel that the simplest theory compatible with available data is always to be preferred, and so we consider the HLA-D antigens to be controlled by genes at one locus, until this is disproven.

3.3. HLA-DR (HLA-D Related) Antigens

In an attempt to establish serological typing for human HLA-D (MLC) determinants, *van Leeuwen et al.* (169) immunized volunteers who produced specific antibodies which, after appropriate absorptions, reacted in a complicated immunofluorescence test with about 20—30% of the peripheral blood lymphocytes of some individuals. The antigen in question was called P1 and was found closely associated but not identical with the HLA-B8 antigen. Later, *van Rood et al.* (246) made extensive absorption of anti-HLA antisera with platelets — which lack HLA-D and HLA-DR antigens — and the absorbed sera could be shown to be cytotoxic against B lymphocyte-enriched suspensions in a modified lymphocytotoxic test. In this way, they defined three different B lymphocyte antigens which are very strongly associated with three HLA-D antigens.

These observations were soon confirmed by others (119), and during the 7th International Histocompatibility Workshop (120) it became clear that it is possible to define a number of B lymphocyte antigens which correspond closely to the HLA-D antigens, and accordingly, they were named HLA-DR (HLA-D related) antigens (table VII). It is still an open question whether HLA-D and DR antigens are identical. In general, the DR antigens seem to be somewhat 'broader' than the corresponding D antigens (6, 120), i.e. the DR antigens are more frequent in the population than the corresponding D antigens, which seem to be 'included' in the DR antigens, e.g. all Dw4-positive individuals are also DRw4-positive while the reverse is not necessarily true. Cells from *DRw4*-homozygous individuals do not always behave as *Dw4*-homozygous typing cells in MLC. *Troup et al.* (327) performed simultaneous D and DR typing of a group of American Indians using Caucasian D-typing and PLT cells and mainly Caucasian anti-DR reagents. They found that some of the D and DR antigens, which are strongly associated in Caucasians, show a similar relationship in Indians while others do not. One explanation for these discrepancies may be that the anti-DR sera are not really monospecific but cross-reacting reagents. Another possibility is, of course, that HLA-D and DR antigens are not truly identical, but controlled by genes at neighboring loci with a high degree of linkage disequilibrium (cf. p. 29). These two possibilities are not mutually exclusive. It is possible — perhaps even likely — that some HLA-D antigens and DR antigens are different determinants on the same molecules. In this context it is worth noting that the HLA-DR antigens are recognized by means of classical immunoglobulin molecules whereas the HLA-D antigens are recognized by the so-called T lymphocyte receptor which is not a typical immunoglobulin.

Independent of the discovery of HLA-DR antigens in man, similar observations were made in animals. When attempts were made to immunize mice with cells from strains differing only for Ir and MLC genes but not for the classical SD antigens (called H-2K and H-2D in that species), it appeared that antibodies were formed which reacted predominantly with the B lymphocytes of the immunizing donor — see *Klein* (152) and *Shreffler and David* (267) for review. The antigens eliciting the antibody formation are called *I region-associated* (Ia) because they are coded for by genes very close to the Ir genes. As illustrated in figure 5 (p. 36), there are several closely linked I loci within the H-2 complex of mice. By analogy, it may be anticipated that additional HLA-DR loci are found in man.

Like HLA-D antigens, HLA-DR and Ia antigens are present on both B lymphocytes, monocytes, macrophages, sperm, and epithelial and endothelial cells (151, 152, 267, 338). The epithelial cells in the skin carrying DR antigens seem to be the so-called cells of Langerhans which are related to macrophages (151). Lymphoblastoid cell lines cultured *in vitro* are almost always of B lymphocyte origin and possess DR antigens, as do lymphocytes from patients with chronic lymphocytic leukemia (CLL). In fact, the first record of these special antigens was made in 1971 by *Walford et al.* (333) who described the so-called 'Merritt' antigen on CLL cells.

3.4. Immune Response (Ir) Genes

As discussed below, the major histocompatibility systems of a number of vertebrates, including mouse and monkeys, have been found to control specific immune-responsiveness against certain antigens (28, 104, 152, 183), and – apart from the pronounced homology between HLA and the corresponding systems in these animals – there is increasing evidence that HLA also contains Ir determinants: (1) the ability to develop immunity towards the E antigen of ragweed seems to be inherited together with certain HLA haplotypes in families (34, 172); (2) the presence of antiadrenal (auto)antibodies in idiopathic Addison's disease is significantly associated with the B8 and Dw3 determinants (312); (3) delicate statistical analyses of antibody titers in families indicate that HLA influences production of antibodies towards various antigens (43); (4) the observation that HLA-D antigens are involved in the cooperation between monocytes and T lymphocytes in the secondary immune response to tuberculin *in vitro* (30, 112) indicates that the HLA system is involved in the immune response; and (5) finally, the fact that nearly all associations so far observed between diseases and HLA primarily concern B or D series determinants, whereas the associations with A and C series antigens are secondary, has been taken as evidence in favor of the existence of Ir genes located between the B and D loci or in the neighborhood of these (cf. section 9.4.).

3.4.1. Function of Immune Response Determinants

From what has been said above, it will be understood that our knowledge of Ir determinants is primarily based on evidence obtained from studies of animals, mice and guinea pigs in particular. However, as there are many reasons to believe

Table VIII. Ir determinants

Animal (genotype)	Antigen			
	Ag1	Ag2	Ag3	Ag4
a/a (C57)	+	−	+	−
b/b (CBA)	−	+	−	−
a/b (F$_1$)	+	+	+	−

+ = (1) Good antibody response (IgG); (2) cell-mediated immunity. Note the inheritance of responsiveness is dominant.

that the situation in humans is similar, and as these determinants may be expected to be in the center of much research in the immediate future, we find it indicated at this juncture to outline the observations made in animals as they appear from recent reviews of *Green* (104), *Benacerraf* (28, 29), *Klein* (152), and *Rosenthal* (248).

The function of Ir genes is perhaps best illustrated by the observations made in mice immunized with two branched multichain synthetic polypeptide antigens as summarized in table VIII. The two antigens are (1) (T,G)-A--L (abbreviated 'Ag-1' in the following) which is composed of short sequences of tyrosine and glutamic acid coupled to *D-L*-alanine side chains on a long backbone of a poly-*L*-lysine polymer, and (2) (H,G)-A--L (abbreviated 'Ag-2') in which histidine has been substituted for tyrosine in the Ag-1 polypeptide. As shown in table VIII, mice of the inbred (homozygous) strain C57 make good antibody responses when immunized with Ag-1 while they respond poorly to Ag-2. In contrast, CBA mice respond poorly to Ag-1 but well to Ag-2. When crosses are made between these two strains, the heterozygous F1 hybrid offspring (carrying all genes from both parents) responds well to both antigens which shows that the ability to make these antibodies is controlled by autosomal *dominant* genes, now called Ir genes. Moreover, when backcrosses are made between the F1 hybrids and each of the parental strains, linkage studies can be made, and it appears that the ability to respond to Ag-1 is inherited in close linkage with the *H-2^b* haplotype of C57 mice while anti-Ag-2 is only formed by animals possessing the *H-2^k* haplotype of the CBA strain. Accordingly, the Ir genes are *closely linked to the H-2 genes*. In fact, it has been found that the Ir genes are located right in the middle of the H-2 gene complex (cf. below).

Similar observations have been made for a variety of other antigens including some, but not all, more complex, naturally occurring antigens, and for other species, e.g. guinea pigs and rhesus monkeys (19, 104).

It seems to be characteristic for most of the antibody responses controlled by H-2 linked Ir genes that cooperation between T and B lymphocytes is needed for the antibody response to occur: these antibodies are so-called 'thymus dependent'. In contrast, antibody responses to antigens of carbohydrate nature do not seem to involve Ir determinants, and these antibodies are 'thymus independent'. The Ir genes not only control antibody responses but also the induction of *cell-mediated immunity* (which is also a T-lymphocyte function) to the antigens in question. Accordingly, there are good reasons to think that Ir genes primarily control the specific functions of T lymphocytes. The nature of the Ir genes products is unknown, but evidence is now accumulating that there is a close relationship between Ia antigens and Ir determinants. This relationship is discussed in more detail in section 8.2.1, but the situation may be briefly summarized here. While it was earlier believed that the Ir genes coded for specific cell-surface antigen receptors on T lymphocytes, there is now an increasing belief that the Ir gene products are mainly expressed on macrophages where they serve a function in the presentation of foreign antigen to immunocompetent T helper lymphocytes (123a). These lymphocytes can help either (i) B lymphocytes to transform into plasma cells producing IgG antibodies to the antigen in question, or (ii) other T lymphocytes involved in cell-mediated immunity. Because the *Ir* and *Ia* genes map so closely within the H-2 system and since both seem to operate at the level of the macrophage, they may in fact be identical: there is no solid evidence against this assumption.

Recently, it has been shown that a part (IJ) of the I region also controls the formation of specific suppressor cells which can suppress the immune response (301). The genes responsible for this action are called *immune suppressor (Is)* genes.

As random, 'unrelated' animal populations have not yet been adequately studied, it is not known whether there are associations between some Ir genes on one side and MLC and SD genes on the other. However, studies of recombinant mice have shown that there are several loci with different Ir genes within H-2 in this species and that they are situated close to the strong MLC locus (cf. p. 36). Extrapolating from the situation of the A, B, C and D series in man, it would be expected that there is association between some of these determinants and specific Ir genes.

3.5. Complement Components, Bf, C2, and C4

Complement is a series of at least ten different factors present in fresh serum (204). The factors are named C1, C2, C3, C4, etc., and the first of them is activated by antibodies which have reacted with their corresponding antigens.

This activated C1 component activates enzymatically the C4 component which in turn activates C2, etc. The final result of these consecutive events is damage to the cell membrane carrying the antigen with which the antibody reacted and this often causes lysis of the cell, e.g. a bacterium (cf. fig. 12, p. 81). In addition, activated complement components have a number of other biological activities such as immune adherence, chemotaxis, and histamine release. Briefly, together with antibody, complement factors are important immune mediators in the body's defence against microbial attacks. The complement system is very complex and recently it has been confirmed that it can be activated not only via C1 (the classical pathway) but also via C3 through the 'alternative' pathway which involves the so-called properdin factors.

Some of the complement components can be traced genetically either by qualitative or by quantitative variations in the population. Properdin factor B, Bf, is polymorphic (with the alleles Bf^F, Bf^S, and Bf^T), and the corresponding locus is located within the HLA complex (8). Lack of C2 has been described as an autosomal recessive trait, and as individuals heterozygous for the deficient gene *(C2O)* have decreased concentration of C2 in serum, it has been possible to do linkage studies for the C2 deficiency gene and it has been found that this gene — and thus most probably also the gene for normal C2 — is closely linked to HLA (99, 131).

By analogy, there is evidence from family studies of C4-deficient individuals that the *C4* gene is also linked to HLA (237), and this has been confirmed by studying the C4 polymorphism (213, 302). As described in the next section, it has very recently been found that there are probably two genes for C4 in the HLA complex and that these are identical to the Chido and Rodgers blood group genes (213). The C1r subunit of the C1 factor is not linked to HLA (72) and the two electrophoretically determined variants of C3 (C3^F and C3^S) are not controlled by HLA-linked genes (165). The linkage relationships between HLA and complement have recently been reviewed by *Lachmann and Hobart* (161) who concluded that C6, C7, and C8 are also not closely linked to HLA.

Thus, in addition to genes controlling antigens (A, B, C, D and DR) and immune response determinants, the HLA complex also contains genes controlling some immune mediators.

3.6. The Bg, Chido, and Rodgers Blood Groups

A notable characteristic of these red cell blood group systems is that they have been difficult to define due to varying 'strength' of antigen on cells from different individuals which has also caused considerable serological troubles in the cross-matching of blood for transfusion (231).

It now appears that the Bg^a antigen is strongly associated with (included in) HLA-B7, Bg^b with B17, and Bg^c with A28 (200, 201). It does not seem quite clear to us whether these antigens have been absorbed from the serum, or whether the agglutinations observed are due to immune-adherence, conglutination or related phenomena. In extensive absorption experiments, *Harris and Zervas* (116) could demonstrate HLA antigens on reticulocytes but not on mature red cells. In any case, it seems reasonable at the present time to consider the Bg system part of the ABC series.

The Chido antigen is present in about 98% of Caucasians, and it is recognized much more reliably in serum than on red blood cells (191, 231). Recently, it appeared that the *Chido* genes are very closely linked to the HLA system: one possible recombination (but not an unequivocal one) has been observed (191). The Rodgers blood group antigen resembles the Chido antigen because it has a high frequency in Caucasians, and it is also primarily present in the serum (231). The *Rodgers* gene is also closely linked to HLA (101b). Very recently, *O'Neill et al.* (213) showed in an elegant series of experiments that the Chido and Rodgers antigens are most probably identical to two different types of the fourth component (C4) of complement. They showed that individuals lacking the electrophoretically fast-moving (C4F) component of C4 are always Rodgers-negative, while those lacking the slow-moving (C4S) component are Chido-negative, whereas the rare individuals lacking detectable C4 are negative for both Rodgers and Chido. As the final proof, they showed that purified C4F and C4S specifically inhibit anti-Rodgers and anti-Chido antisera, respectively (213). Thus, the Chido and Rodgers blood group substances are identical to the C4S and C4F proteins, respectively, and because there is no allelic relationship between the *Rodgers* and *Chido* genes, there are probably two different *C4* genes in the HLA region.

3.7. Genetic Relations between HLA Determinants

Although the genes of the A, B, C, and D series are closely linked, a number of families showing crossing-over between these series have been reported (289). Most information exists in relation to the A and B series. The latest combined study (26) concerned 4,614 meiotic divisions in which 40 crossing-overs had occurred. Thus, the recombination frequency between the A and B series is 40/4,614 = 0.0087 or 0.87 centimorgans with a standard error of 0.14 centimorgan. The distance between the B and D loci is about 0.5 centimorgans (142, 315). Ten A-B recombinants informative for C series antigens have been reported (cf. 211); in eight of these, the C antigens followed B, while the A and C antigens travelled together in the remaining two. Thus, the distance between

Table IX. HLA-A,B haplotype frequencies × 10^3 in Danes

	A1	A2	A3	A9	A25	A26	A11	A19	A29	A32	A28	Gene frequency
B 5	3	22	6	5	1	1	4	7	<1	1	5	55
B 7	9	52	*54*	12	<1	1	5	10	1	1	1	142
B 8	*98*	4	6	4	6	5	1	4	<1	2	6	127
B12	13	*63*	6	16	2	<1	4	25	*15*	8	10	136
B13	1	6	<1	2	<1	2	<1	7	<1	2	<1	22
B14	1	3	5	2	1	1	1	5	1	1	5	23
B15	6	*47*	14	11	<1	1	3	<1	2	4	9	94
B17	*15*	17	3	1	<1	<1	<1	4	2	1	1	39
B18	2	9	2	3	*5*	3	2	1	1	1	6	36
B21	1	8	2	3	1	1	1	<1	1	<1	2	18
B22	1	7	3	1	1	4	3	<1	<1	3	2	19
B27	<1	15	6	6	<1	4	3	5	<1	3	3	44
B35	2	9	*25*	9	3	1	*15*	3	<1	1	5	72
B37	3	2	<1	1	<1	<1	<1	1	<1	<1	<1	6
B38	4	<1	1	2	<1	*8*	1	<1	<1	1	4	13
B39	<1	8	<1	3	<1	<1	3	3	1	1	<1	14
B40	6	*50*	7	6	1	1	7	12	1	2	6	95
B41	<1	<1	<1	<1	<1	2	1	3	<1	2	2	7
Gene frequency	168	323	145	92	19	30	52	99	23	31	51	

The frequencies have been multiplied by 10^3 and are from *Staub Nielsen et al.* (210). Figures in *italics* indicate haplotypes with significantly (p <0.001) positive Δ-values. The haplotypes involving *A10* and *B16* are not given as these determinants are the sums of *A25* and *A26*, and *B38* and *B39*, respectively. *A29* and *A32* are both completely included in *A19*, but this determinant includes also other determinants (*A30*, *A31* and *A33*), which were not investigated in this study.

the A and C locus can be estimated to be about 4/5 × 0.87 = 0.7 centimorgans, and that between the C and B about 0.2 centimorgans.

Despite the fact that crossing-over is known to take place between the various HLA loci, some *HLA* genes tend to occur significantly more frequently in the same haplotype than should be expected under the assumption of linkage equilibrium. Table IX shows the frequencies of the pairwise haplotypes of the *HLA-A* and *B* alleles in a northern Caucasian population (210). One of the most well-known examples of this *linkage disequilibrium* or *gametic association* (cf. below) concerns the *A1,B8* haplotype which occurs about five times as frequent-

Table X. Linkage disequilibrium (gametic association)

Haplotype			Frequency, %	
A	B	D	observed	expected
A1	B8		9.8	2.1
A3	B7		5.4	2.1
	B8	D3	8.6	1.4
	B7	D2	3.9	1.8

The expected haplotype frequencies were calculated under the assumption of no association.

ly as should be expected (table X). Obviously, this association at the haplotype level is reflected in a strong association between the A1 and B8 antigens at the population level (table XII): about 81% of all B8 positives also possess the A1 antigen although this antigen has a frequency of only 31% in the random population.

Table XI shows the pairwise haplotype frequencies and Δ-values (see below) for the most positively associated combinations *A-B, C-A, C-B* and *D-B* (134, 210, 309). Almost all factors from a series are strongly associated with one or more factors from each of the other series although the strength of the associations decreases with increasing distance between the corresponding loci; thus the strongest associations exist between the C and the B series which are also the most tightly linked. Too few random individuals have yet been typed for MLC to detect associations for all MLC antigens.

As mentioned earlier, the DR antigens seem to be very strongly associated with the MLC determinants. One problem of considerable interest from a genetic point of view concerns the possibility of whether three or more closely linked HLA genes may occur more frequently in the same haplotype than should be expected from the pairwise associations between these genes. In general, this does not seem to be the case, but there may be at least one exception: the *A23,Cw4,B12* haplotype is by far the most frequent (perhaps the only) haplotype carrying *Cw4* as well as *B12,* and the term *'superhaplotype'* has been suggested (210) to indicate such haplotypes in which three or more alleles are more strongly associated than expected from the pairwise associations.

The two most common variants of the properdin factor Bf both show associations with the HLA-B and D-series: factor Bf^F is associated with B12, and Fb^S with B35 and D1 (8; *Rubinstein,* personal commun.). The so-called *C2^O*

Table XI. Frequencies and Δ-values × 10^3 of some *HLA-A,B, C,A, C,B,* and *D,B* haplotypes in Caucasians

Combination	Haplotype	Frequency	Δ	Combination	Haplotype	Frequency	Δ
A,B	A1,B8	98	77	C,A	C1,A26	7	6
	A1,B17	15	8		C2,A2	33	17
	A2,B12	63	19		C3,A2	91	28
	A2,B15	47	16		C4,A3	28	15
	A2,B40	50	19		C4,A11	15	10
	A3,B7	54	33	C,B	C1,B22	16	15
	A3,B35	25	14		C1,B27	17	16
	A11,B22	3	2		C2,B27	31	29
	A11,B35	15	11		C2,B40	14	10
	A25,B18	5	5		C3,B15	96	77
	A26,B38	8	7		C3,B40	86	68
	A28,B14	5	3		C4,B35	69	63
	A28,B15	9	3	D,B	D1,B35	33	27
	A28,B18	6	4		D2,B7	39	26
	A29,B12	15	11		D3,B8	86	72
	A29,B45*	2	2		D4,B15	43	35
	A32,B41	2	2		D5,B16	15	13
					D6,B16	8	7
					D7,B13	9	8
					D8,B40	13	9

The values for the *A,B, C,A,* and *C,B* haplotypes are from *Staub Nielsen et al.* (210), and those for the *D,B* haplotypes are from *Thomsen et al.* (309) except for the *D5,B16* and *D6,B16* haplotypes which are from a joint report (134). Most of the Δ-values are very significantly different from zero.

gene, present in double dose in individuals lacking the C2 component of complement, is very strongly associated both with the *B18* and the *D2* genes; almost all C2-deficient persons are *D2* homozygous and many are also *B18* homozygous (99, 100). As *D2* is normally more strongly associated with *B7* than with *B18,* the *B18,D2,C2^O* haplotype may be considered a superhaplotype (131). It has been assumed that the *C2^O* gene arose by mutation on an *A25,B18,Dw2* haplotype and that it has partly become separated from *A25* and *B18* in some haplotypes due to crossing-over. Under this assumption, it can be estimated that the mutation may have taken place about 1,650 years (55 generations) ago (297).

Chido-negativity (absence of C4S) is associated with HLA-B12 and B35 and Rodgers-negativity (absence of C4F) is associated with HLA-B8 (87).

In conclusion, there appears to be linkage disequilibrium between all HLA alleles recognized so far, and this is, in fact, a major reason for considering HLA as one genetic system. In contrast, there is no association between the *GLO* alleles (cf. p. 31) and any HLA antigens (113).

Table XII. The 2 × 2 table examplified with the association between the HLA-A1 and B8 antigens in 1,967 unrelated Danes

	Number of individuals		Total
	B8-positive	B8-negative	
A1-positive	376 (61.5%)	235	611 (31.1%)
A1-negative	91 (6.7%)	1,265	1,356
Total	467 (23.7%)	1,500	1,967

The table is often given as follows:

First antigen	Second antigen	+/+ a	+/− b	−/+ c	−/− d	Total N
A1	B8	376	235	91	1,265	1,967

where, for example, +/− means number of individuals possessing the first character (A1) and lacking the second (B8).

The *chi-square* is

$$\chi^2 = \frac{(ad-bc)^2 N}{(a+b)(c+d)(a+c)(b+d)} = 699.4$$

corresponding to a *correlation coefficient*

$$r = \sqrt{\chi^2/N} = \sqrt{699.4/1967} = 0.60.$$

Gene frequencies for *A1* and *B8* can be calculated by *Bernstein's formula*

$p = 1 - \sqrt{1 - a}$ (where a is the antigen frequency) to 0.170 and 0.127, respectively.

The Δ-*value-* can be calculated by the formula

$$\Delta = \sqrt{\frac{d}{N}} - \sqrt{\frac{(b+d)(c+d)}{N^2}} = 0.077.$$

Thus the frequency of the *HLA-A1,B8 haplotype* is $h_{A1,B8} = P_{A1}P_{B8} + \Delta_{A1,B8} = 0.170 \times 0.127 + 0.077 = 0.099$.

3.7.1. Association

Studies of association between various characters play a major role in much HLA-related research, and accordingly, we find it relevant to discuss this topic here. We shall, however, restrict ourselves almost exclusively to pairwise associations as these are the most easy to handle. Detailed mathematical and statistical descriptions can be found in *Armitage* (13) and *Li* (174).

Pairwise associations are investigated by *2 X 2 contingency tables* as illustrated in table XII. Distinction should be made between the strength of an association and its statistical significance. In general, the strength reflects biological importance. An association may well be strong but statistically insignificant (when the number of items investigated is small), and a weak association may become highly statistically significant when many items are studied.

Statistical significance can be measured by Fisher's exact test or by various approximate tests, usually the χ^2 test. When one or more of the *expected* entries in the 2 X 2 table are less than five, *Fisher's exact test* is usually the only reliable one. By means of computers, this test can be applied also to large samples, but, in these cases, the χ^2 *test* can be used with sufficient accuracy. Occasionally, *Yates' correction* for discontinuity is used in the χ^2 test, and it is worth noting that this does not correct for small sample size, and that it should not be used when a number of χ^2 values are to be added.

The *strength* of an association is independent of sample size and can be estimated in various ways, the two most common being the coefficient of correlation and the relative risk. The *coefficient of correlation* (r) is estimated by

$$r = \sqrt{\frac{\overline{\chi^2}}{N}}$$

where N is the total number of patients and controls. This coefficient varies from zero (no association) to one (absolute association, i.e. the factors are always present together, the entries b and c being zero); calculated in this way, it is always positive — also for negative associations, but it can arbitrarily be given a negative sign for negative associations. In this relation it is worth noting that negative associations are just as important from a biological point of view as are positive ones. The *relative risk,* primarily used for the associations between HLA and disease, is discussed on page 67.

3.7.2. Haplotype Frequencies and Δ-Values (49, 52, 174)

If two genes on two linked loci are in *linkage equilibrium,* i.e. if there is no association between them, the frequency of their occurrence together on the same chromosome must be the product, p_1p_2, of the gene frequencies, p_1 and p_2, i.e. the haplotype frequency, h, would be equal to p_1p_2. If there is *linkage disequilibrium,* the haplotype frequency deviates from the expected p_1p_2 product and the deviation is expressed by the so-called Δ-*value,* which is accordingly defined as $\Delta = h-p_1p_2$. Positively associated factors (e.g. *A1* and *B8*) show positive deltas, while negative associations give negative deltas. Haplotype and gene frequencies can be estimated from family and population studies. In family studies, the haplotypes can usually but not always be ascertained and counted in a sample of unrelated parents (291). The number of the various genes can also be counted and the Δ-values are then estimated from the haplotype and gene frequencies by the above formula. In random-mating populations (i.e. populations in Hardy-Weinberg equilibrium), Δ-values can be estimated from phenotype frequencies of unrelated individuals by a special formula (table XII) (52, 180), and when the gene frequencies are also estimated — e.g. by gene counting or by Bernstein's formula (49, 174); cf. table XII — haplotype frequencies can be obtained from such data by the formula $h = p_1 p_2 + \Delta$. The test as to whether the Δ-value differs significantly from zero is performed as a 2×2 χ^2 test for the association between the two corresponding characters (e.g. antigens) in the population.

The mathematical analysis of associations between three or more genes at three or more linked loci is a much more complicated affair, in particular significance testing is difficult. *Piazza* (226) reported preliminary formulas for the three-locus problem.

4. Linkage Relationships of HLA

The first linkage between HLA and another genetic system was found by *Lamm et al.* (164) who observed that the locus for PGM_3 (phosphoglucomutase-3) is linked to HLA with a distance of about 15 centimorgans in males and 30 in females reflecting the well-known fact that recombinations are more frequent in females than in males. By studying families with HLA recombinants, *Lamm et al.* (163) later found strong support for the assumption that the PGM_3 locus is closer to the B than to the A locus of HLA. More recently, it has been found that the locus for glyoxylase (GLO) is about 5 centimorgans from the HLA-B locus (113, 173, 334). As PGM_3 is linked to the P blood group system and a locus for urinary pepsinogen (186), these five genetic systems (HLA, GLO, PGM_3, P, and Pg — urinary pepsinogen) are controlled by genes carried by the same autosome.

Experiments with cell hybrids between chinese hamster and human cells have yielded strong evidence that the PGM_3 locus is present on chromosome No. 6 (280), and further strong support for this observation was obtained by *Lamm et al.* (162) who found that HLA segregated together with a large pericentric inversion on this chromosome in a large kindred. This study showed that HLA is most likely to be within 75 centimorgans from the centromere.

Recently, *Francke and Pellegrino* (98) studied man-hamster somatic cell hybrids which contained defined parts of human chromosome No. 6. These were obtained from fibroblasts carrying a balanced reciprocal translocation between the short arms of chromosomes No. 1 and 6. Their results indicated that MHC is located on the short arm of No. 6 (at 6p21), about 25 centimorgans from the centromere.

The discovery of very close linkage between HLA-B and the locus for a gene leading to deficiency of the enzyme 21-hydroxylase, made recently by *Dupont et al.* (86, 87) may have wide implications on theoretical as well as practical work and thinking in this area. Later additions and studies by others have until now revealed no crossing-over between these two loci. It is, however, worth noting that association may exist between some HLA-B antigens and this enzyme deficiency (106), which raises the question whether the gene controlling 21-hydroxylase production belongs to the HLA system, a problem which will not be too easy to decide.

Snell (275) first drew attention to the observation that some animal species, e.g. man and cattle, have two complex blood group systems: one on red cells and one on leukocytes, while others, e.g., mice and chickens, have one complex system controlling antigens on both kinds of cells. In this relation, it is an interesting fact that the factors of the HLA and PGM_3 systems are mostly confined to nucleated cells, perhaps leukocytes in particular, while those of the related erythrocyte systems, Rhesus and PGM_1, are present in red cells and also controlled by linked genes, on chromosome No. 1 in this case. Thus, the HLA and Rhesus systems might be evolutionarily related.

5. Geographical Variation

The purposes of studying genetic systems like HLA in various populations are among others (1) to clarify the genetics of the system into more detail; (2) to unravel the evolutionary relationship between the populations, and (3) to investigate the extent to which the system is subject to natural selection.

Studies of HLA in a great variety of populations were the main aim of the 5th International Histocompatibility Testing Workshop in 1972 when 29 laboratories typed 54 different populations from many parts of the world with the same set of antisera (118, 133). It appeared that many of the HLA antigens which were well-defined in Caucasians were not so in other ethnic groups. In particular, several sets of sera reacting identically in Caucasian cell panels failed to do so in other populations, and it became clear that some HLA antigens are present only in some populations. An illustrative analysis of the workshop data concerning the efficiency with which various HLA antigens could be defined on a worldwide basis has recently been reported by *Bodmer et al.* (36).

Piazza et al. (228) used the same data and constructed 'evolutionary trees' based on the differences in gene frequencies between populations for HLA and other polymorphic systems. In general, these trees supported previous theories on the evolutionary relationship between populations and added significantly to this knowledge.

Bodmer et al. (38) analyzed the workshop data by three different genetic methods previously developed to provide information as to whether the HLA system is subject to natural selection, and each of these methods gave evidence that this, in fact, seems to be the case.

More recently, *Ryder et al.* (252) tabulated data on HLA-A and B antigen frequencies in about 50 population samples from various parts of Europe. Most of these frequencies varied considerably from one place to another. For example, it can be seen from figure 4 that the frequency of B8 varies from below 10% in parts of the Mediterranean to above 30% in Scotland. These authors also calculated genetic distances between most of the populations and found the smaller distances the closer the populations lived.

In the following, we shall give a very brief description of the general HLA characteristics of some representative population groups. Only the A and B antigens have been sufficiently studied, and thus we shall limit the survey to deal

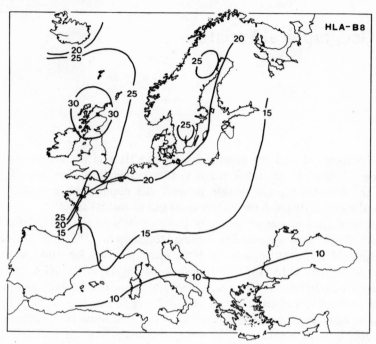

Fig. 4. Variations in the frequency of the HLA-B8 antigen within Europe. The lines indicate 'boundaries' between areas having approximately the same frequency. Figures are antigen frequencies in percent. From *Ryder et al.* (252).

with these. For more detailed information the reader is referred to the joint reports (133, 135, 252).

A characteristic feature of *Caucasians* is the common occurrence of the A1 and B8 antigens which are very infrequent or absent in most other groups. In Caucasians, there is moreover a very strong gametic association between *A1* and *B8*. Another typical Caucasian haplotype is *A3,B7*. However, it is worth noting that these haplotypes are European rather than Caucasian, and the corresponding antigens decline in frequency from the North to the South of Europe. In contrast, B5 and Bw35 increase in that direction. In the Middle East, where Bw21 has a high frequency, and in India, the *A1,B8* and *A3,B7* haplotypes are no longer frequent.

Two antigens, Aw36 and Bw42, seem only to exist in *African Negroes*, in whom Bw17 is the most frequent antigen. Nevertheless, most well-known Caucasian HLA antigens including A1, A3, B7, and B8 are present in Black Africans.

In *Mongoloids,* high frequencies of A9, Bw40, and Bw22 are characteristic. A1, A3, B7, and B8 are absent in most Mongoloid populations. In Eskimoes, A9 is particularly frequent, and there seems to be an 'Eskimo-specific' antigen, Bw48, in the B segregant series.

American Indians have few of the well-known HLA antigens but as 'blanks' are not frequent, there must be a high degree of homozygosity for the few antigens present, A2, A9, Bw35, Bw40, and Bw15 in particular. Bw15 is primarily present in South American Indians, while B27 is primarily a North American allele.

The A2 antigen has appreciable frequencies in almost all populations studied and reaches its highest occurrence (about 80%) in American Indians.

Finally, it seems worth noting that some HLA antigens are recognizable by human alloantisera in chimpanzees: A1, A11, Bw15, Bw17, and Bw22, while most other antigens — including A2 — are probably absent.

6. Homologous Systems in Animals

When HLA was recognized as the major histocompatibility system (MHS) in man, a similar system, H-2 had already been known for years in mice (152). Since then, a number of other vertebrates has been studied, and in everyone of these species an MHS has been found. The names of these systems are H-2 in mice, AgB in rats, B in chickens, SL-A in pigs, DL-A in dogs, RhLA in rhesus monkeys, and ChL-A in chimpanzees. Moreover, all these systems seem to control related characters such as 'classical' serologically detectable antigens, MLC-activating antigens, immune response determinants, and immune associated antigens whenever these have been looked for.

Figure 5 shows the arrangements of genes within the H-2 and RhLA systems (19, 153, 239). The former has been mapped most accurately because mice are bred so easily. The most marked difference between the H-2 and the HLA system is the position of the genes controlling the strong MLC antigens: whereas these genes are located outside the A and B loci in man, they are between the corresponding loci, H-2D and H-2K in mice. In rhesus monkeys, however, the strong MLC locus is located outside the 'classical' loci corresponding to the human situation. Moreover, this seems also to be true in rats (45), and thus the arrangement in mice may be an extraordinary exception. Most strikingly, the Ia loci are in all three species located close to the MLC locus, and while the Ir loci have not yet been mapped in man, they are also close to MLC both in mice and in rhesus monkeys. These observations support the concept that Ia, Ir, and MCL are related in their function. Moreover, an analogue of the Bf locus has recently been mapped close to the MLC locus in rhesus monkeys (345), as in man. The alleles of the Ss and Slp loci (locus?) of mice control quantitative variation and partially sex-limited properties of one or more serum proteins, partly or totally identical to the fourth component (C4) of complement (59, 160). A homologue of the human C locus has not been clearly identified in mice, and man is the only species in which linkage disequilibrium has been extensively studied because it is the most outbred species. However, linkage disequilibrium seems also to be pronounced in the B system in chickens (*Simonsen*, personal commun.). Nevertheless, many more genetic details are known on the H-2 system than on the HLA system; recent reviews of *Klein* (152), *Snell et al.* (277), and *Shreffler and David* (267) illustrate the complexity of H-2. One of

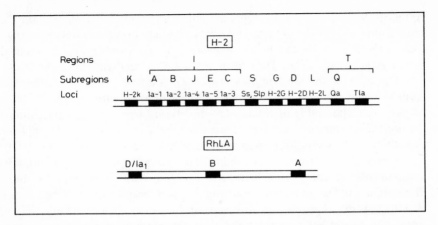

Fig. 5. The genetic arrangement of the major histocompatibility systems in mice (H-2) and Rhesus monkeys (RhLA). Modified from *Klein et al.* (153) and *Roger et al.* (239). The H-2 system belongs to the ninth linkage group and is on chromosome 17. The K and D regions correspond to the A and B series of HLA, and the L region may correspond to HLA-C. The I region contains genes controlling Ir, Ia, and MLC determinants. The Ss, Slp locus control C4 and sex-limited protein. It is uncertain whether the T region belongs to H-2. In the RhLA system, the A and B loci correspond to the HLA-A and B loci, and D and Ia₁ to the HLA-D and DR loci, respectively. Ir and Bf have been mapped close to the D locus.

the most striking observations is that the I (immune) region of H-2 can be divided into several subregions (loci?) because various characters controlled by this region have been separated by crossing-over in hybrids between different congenic strains which differ only at the H-2 complex. The characters controlled by this region are specific immune responses towards various foreign antigens, Ia antigens, MLC stimulation, and gvh reactivity. In most cases, a high immune response dominates over low responses, but recently a dominant low response has been described and shown to be due to the formation of suppressor cells (29). Some *immune suppressor (Is)* genes have been mapped to the IJ region.

The H-2 complex does not only control various immune reactions, but also a variety of other quantitative traits as recently reviewed by *Ivanyi* (124). For example, the weights of the testis and the thymus are influenced by H-2 as is the concentration of testosterone in the serum and of cyclic AMP in the liver. There even seems to be a certain mating preference related to the H-2 system (342).

Before leaving the H-2 system, we would like to mention two H-2 linked systems as they may be of relevance for future human studies.

The first of these is the thymus leukemia antigen (Tla) system which codes for alloantigens on thymocytes but apparently not on mature T lymphocytes;

strangely enough, even mice who do not carry a given Tla antigen on their normal thymocytes can develop leukemias with cells expressing that Tla antigen. This is thought to be due to an interaction between oncogenic virus and the mouse genome (152). The Tla locus is about 1.5 centimorgans from the D locus.

The other is the T-t complex (14, 152), which is about 14 centimorgans from H-2, and contains a considerable number of genes with a variety of effects ranging from shortening of the tail to lethality *in utero*. The two extraordinary features of the various recessive *(t)* lethals of the T-t complex are (1) a dramatic distortion of the segregation ratio in +/*t* male X +/+ female matings (+ indicating the normal alleles): too many or too few (depending on the *t* allele) +/*t* offspring are observed in comparison with the expected 50%. This may be due to differential survival of sperms carrying + and *t,* respectively; (2) the lethal *t* alleles markedly suppress the crossing-over frequency in the vicinity of the T locus, and this effect goes right along the chromosome all the way between T and H-2. In fact, there seems to be linkage disequilibrium between H-2 and t complexes (110). There is increasing evidence that the T-t complex codes for differentiation antigens present on sperm and involved in some stages of fetal development. A homology between T-t and H-2 has been suggested (14). In this connexion, it may perhaps be noted that attempts to demonstrate a relationship between HLA and spina bifida in man have so far been inconclusive (10, 35).

7. Notes on the Biochemistry of HLA Antigens

The biochemical characterization of the HLA antigens has met with considerable difficulties, but it now seems clear that the HLA antigens (as well as their homologues in other species investigated) are of glycoprotein nature, and that the alloantigenic variation resides in the polypeptide moiety and not in the carbohydrate side chains (218). The difficulties stem from the fact that the concentration of HLA antigenic substance in serum and other body fluids is very low and thus it has been necessary to solubilize or extract the antigens from other sources such as peripheral blood leukocytes, spleen lymphocytes, or from cultured cell lines with known HLA constitution. The solubilization procedures have varied from high concentration of chaotropic ions (e.g. $3\,M$ KCl) to enzymatic digestion (e.g. by papain), detergent treatment (e.g. sodium dodecyl sulfate) and sonication, and even extraction with ether/alcohol mixtures has been employed (117, 207, 234, 322). Many of the conflicts in the early literature appear to originate from differences in the initial solubilization steps used.

The later steps in the procedure have generally been attempts to purify and concentrate the first obtained material by the available methods, such as the various chromatographic and electrophoretic techniques, ultracentrifugation and lyophilization. The precise position of the antigenic substance during the fractionations is usually followed either by assaying the potency of a given fraction to inhibit a known relevant serologic typing reaction, or by sensitive indirect precipitation assays employing specific anti-HLA sera and anti-human immunoglobulin combined with radioactive labelling of the membrane components. Often the information obtained at this stage has been sufficient to give a rough estimate of the molecular weight and subunit structure and in only relatively few studies has further characterization of amino acid composition or sequence and carbohydrate content been carried out.

The picture emerging from these studies shows that the HLA antigens apparently fall into two distinct structural (and probably functional, cf. later) categories: (1) the HLA-ABC antigens and (2) the HLA-D/DR antigens.

The *ABC antigens* are composed of two noncovalently linked polypeptide chains (fig. 6); a light chain identical to β_2-microglobulin (MW 11,600) and a heavy chain (MW 43,000–45,000) which carries the alloantigenic determinants as well as the carbohydrate (3—8%) (20, 218). Near the C-terminus of the heavy chain is a hydrophobic region which provides an anchorage site where the polypeptide penetrates the cell membrane with a short tail pointing towards the interior of the cell, while the main part (including alloantigenic determinants, carbohydrate and binding sites for β_2-microglobulin) is presented to the surroundings (20, 274, 283).

Information on the primary structure of the heavy chain has only been published for less than the first 20 N-terminal amino acid residues at the time of writing. There appears to be a close homology between at least the HLA-A and B antigens investigated (20), which supports the suggestion that the HLA-A, B and C loci arose as duplications of a common ancestral gene (37). The limited data available is not yet sufficient to prove or rule out the proposed homology between the HLA antigens and the immunoglobulins (37, 224).

The light chain, β_2-microglobulin, is a recently discovered protein (MW 11,600) of unknown function; it is present in low amounts in the serum of all individuals studied and present as a cell surface component of most if not all nucleated cells (324). Patients with certain kidney disorders have high concentrations of β_2-microglobulin in their urine. Recently it has come into the focus of immunology because its structure shows a remarkable homology with especially the third heavy chain domain of the immunoglobulins. It is worth mentioning that the immunoglobulins are not associated with the β_2, that no allotypic variation in β_2 has been reported and that the genes coding for β_2 reside on chromosome No. 15 while HLA-genes are on No. 6 (102).

The arrangement of the polypeptide chains explains some of the early discrepant results, because enzymatic digestion cleaves the HLA-ABC heavy chain at a point close to the cell membrane, thus yielding one soluble part carrying the antigenic determinant and one part remaining embedded in the membrane, while detergent procedures release more or less intact molecules of larger size, and salt extraction causes, at least in some cases, autolytic degradation giving mixtures of both categories (178).

The *HLA-DR antigens* (fig. 6) have recently been shown to be also composed of two polypeptide chains of unequal size (MW 28,000 and 35,000) but both chains are different from those found in the HLA-ABC series antigens and from β_2-microglobulin (20, 150). The data indicates that the HLA-DR specificity is carried by the heavy chain, and the gene for this must therefore be located on chromosome No. 6. The location of the gene coding for the light HLA-DR subunit remains unknown.

There are of course limitations to what these studies disclose about the native configuration of the HLA antigens on the cell surface, but work on the redistribution of surface markers caused by antibodies and assayed by fluorescent indicators has given further important knowledge (31, 156, 263, 324). This phenomenon is known as capping. When lymphocytes or other cells are incubat-

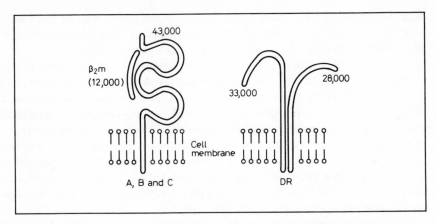

Fig. 6. Tentative models of the HLA-A, B, or C and DR molecules on the cell surface. Modified from *Barnstable et al.* (20). The figures indicate the molecular weights of the polypeptide chains. β_2m = β_2-microglobulin. Within each molecule, the two chains are bound noncovalently together. The chains controlled by HLA genes have molecular weights of 43,000 and 33,000 for HLA-ABC and DR antigens, respectively. The largest parts of the molecules are on the outer part of the membrane and carry the HLA antigenic specificity.

ed with divalent antibodies directed against a surface antigen present on the cell (in the absence of complement which, if present, would cause cell lysis), microaggregates of antigen-antibody complexes form on the surface. These complexes can be induced to aggregate into a single cap upon the application of one more layer of antibody, this time heteroantibody directed against the immunoglobulins first employed. The anti-immunoglobulins in the outermost layer of this 'sandwich' are conveniently used as fluorochrome-labelled antibodies so that the reaction can be followed under the fluorescence microscope. If the antigen investigated is the native immunoglobulin of the B cells, a one-step procedure is sufficient to induce capping. This capping process is not an entirely passive process as it requires an intact energy metabolism and is not restricted to antigen-antibody systems, but is also seen with other receptors and ligands. The cap formed in this way will, under suitable conditions be internalized by endocytosis, leaving a cell with almost none of the original surface antigens left, a phenomenon called 'stripping'. Fresh antigens reappear some hours later if the cells are kept under cell culture conditions.

Studies performed with these two techniques (capping and 'stripping') have shown the following points:

(1) Two different antigens belonging to the same segregant series will, when present on the same cell (e.g. from an *HLA-A1/2* heterozygous individual) form caps and 'strip' independently of one another; the HLA-A1 antigens can be

capped without capping HLA-A2 by using a monospecific anti-HLA-A1 in the first step described above, and vice versa. This illustrates that the products of the allelic genes are independent molecules on the cell surface (31).

(2) Two antigens belonging to different segregant series will also cap independently when present on the same cell (e.g. HLA-A1 and B8) whether or not they are coded for by genes in the same haplotype (31). This demonstrates that the antigenic determinants of the haplotype are situated on distinct molecules. Prior to the formal genetic proof of a separate HLA-C locus by the observation of a crossing-over, this kind of experiments provided the best evidence in favor of the third HLA locus: capping and 'stripping' of the HLA-C antigens do not influence the other HLA antigens and vice versa (278). On the other hand, it has been found that the HLA-Bw4 and HLA-Bw6 antigenic determinants are present on the same molecules as the B antigens (236).

(3) Cap formation of all the HLA-ABC antigens will simultaneously bring about some but not total co-capping of the β_2-microglobulin molecules present on the cell, showing that at least some of the HLA molecules are associated with some of the β_2-molecules (324).

(4) Total capping of β_2-microglobulins gives total co-capping of all HLA antigens which indicates that all HLA molecules carry β_2-microglobulin (324).

The very recent discovery that the histocompatibility antigens play a crucial role in the cellular interactions leading to an immune response to T-dependent antigens (cf. section 8.2.1) raises some extremely important but difficult questions to the biochemists: (i) do the HLA molecules possess special physico-chemical properties to which this function can be ascribed, such as 'stickiness' or special affinity for foreign substances; (ii) is there a variation in this supposed affinity?, and if so, (iii) does this variation coincide with the allogenic variation? The answers to these questions have important implications for understanding the functions of the histocompatibility systems in the immune response and may help explaining some of the HLA and disease associations. Some experimental evidence compatible with these assumptions has been obtained in the mouse H-2 system where the antigens induced by certain murine leukemia viruses are selectively associated with some of the H-2 antigens and not with others (42). In addition, other experiments have shown that H-2 antigens co-cap with a variety of independent cell membrane antigens as TL, Thy-1 and T-200 (40).

It has been pointed out (54) that the physicochemical interactions between molecules confined to move while anchored in the cell membrane may be much stronger than expected if they were in free solution, because the effective concentration is manyfold higher. This may provide a physical explanation for the antigen presentation in general, but the question of possible specificity remains to be answered.

8. Biology of the HLA System

It is naturally a great challenge for all biologists and geneticists involved in the research on HLA in man or related systems in animals to clarify the biological functions of these systems. As apparent from section 3 it is clear that these systems control antigens, immune response determinants, as well as certain immune mediators. It is not difficult to understand that the immune response determinants and the complement components must be of great importance in the body's defence against microbial invasions, but the biological role of the antigens we have been able to determine is much less clear. The great mystery with these antigens is their tremendous polymorphism. How has this enormous degree of diversity been maintained throughout evolution? It seems to us that a key solution to this problem must answer the question: why are we all so different? Before discussing some of the explanations which have been put forward in this context we shall briefly summarize some of the properties which are known to be controlled by these systems. Although most of this knowledge derives from animal studies, there are many reasons to believe that HLA serves similar functions in man.

8.1. Properties of HLA and Homologous Systems

(1) The alloantigens controlled by these systems are, of course, the most well known, and as discussed in section 3, they fall in three groups: the ABC antigens, which are present on all nucleated cells and which readily give rise to circulating alloantibodies upon immunization; the DR antigens primarily present on B lymphocytes and monocytes, and the D antigens present on the same cells and responsible for stimulation in allogeneic mixtures of lymphocytes. As mentioned earlier, DR and D antigens may be identical.

(2) Various specific Ir determinants within these systems are responsible for the development of specific cell-mediated immunity and production of IgG antibodies directed towards the antigens in question. It should be stressed again that HLA linked Ir determinants are necessary for the IgG antibody production against some antigens (T-cell-dependent antibodies), and that the genes coding for the IgG antibodies themselves are not linked to HLA.

(3) The complement components (C2, Bf, and C4) controlled by HLA genes may be considered crucial for the eradication of microbial organisms, and lack of the C2 component seems also to be associated with autoimmune phenomena (e.g. discoid or systemic lupus erythematosus and glomerulonephritis) indicating that an intact complement cascade is one important factor in the protection against such disorders.

(4) As reviewed by *Ivanyi* (124), *Ivanyi and Forejt* (125), *Demant* (75), and *Klein* (152), the H-2 system of mice also influences other apparently nonimmunological functions such as testis and thymus weight, testosterone concentration in the blood, and the amount of cyclic AMP in liver cells is also H-2-influenced (190).

8.2. Biological Function

The theories explaining the biological function and the polymorphism of the HLA system must take into account the above phenomena. Several hypotheses have been put forward (37, 47, 128, 275), but few − if any − of them explain all facets. However, it is conveivable that this system has more than one function. Below we shall discuss some of the hypotheses which have been put forward to explain this problem.

8.2.1. Role of MHC in the Cooperation between Cells in the Immune Response

Since the first edition of this survey, there has been a major breakthrough in our understanding of the biological function of the major histocompatibility complex (MHC). This knowledge derives mainly from experimental studies in animals, and the milestones in this development may be summarized as follows: (i) the discovery by *McDevitt and Benacerraf* (183) of the immune response (Ir) determinants and the demonstration that these determinants belong to the MHC; (ii) the observation of *Doherty and Zinkernagel* (80), *Zinkernagel and Doherty* (347) and of *Shearer* (264) that the mouse H-2D and H-2K antigens (corresponding to the ABC antigens in man) play a fundamental role in the lysis of virus-infected or hapten-conjugated cells; (iii) the experiments of *Rosenthal* (248), *Rosenthal and Shevach* (249) and *Shevach and Rosenthal* (266) who showed that I-region gene products (related to HLA-D/DR gene products in man) are involved in the cooperation between macrophages and T lymphocytes, which is of crucial importance in many immune responses, and finally, (iv) the very recent discovery of *Zinkernagel et al.* (346) that the thymus is the organ where T lymphocytes learn to cooperate with cells of a given MHC type. Clearly, many other investigators have contributed very significantly to our present knowledge, but limitations of space do not allow them all to be acknowledged. For more detailed information the reader is referred to the various reviews (29, 193, 248, 276) and to some references (54, 122b, 123a, 123b, 341, 346). In this review, we only attempt to give a brief account of the current state of knowledge.

The key cells in the immune system are the lymphocytes which may be divided into two major groups: the T lymphocytes which need a functional thymus for their development, and the B lymphocytes which mature in the bursa fabricii in birds but in an unknown place (the fetal liver?) in mammals. The HLA system is involved in *thymus-dependent immunity* which comprises cell-mediated immunity and production of IgG antibodies towards a variety of antigens (thymus-dependent antigens). Thymus-dependent immunity is of major importance in the defence of the organism against viral and fungal infection, in graft rejection, and in contact dermatitis. The central cell in thymus-dependent immunity is the *T lymphocyte*, but in its action it requires cooperation with two other cell types: macrophages and B lymphocytes. Moreover, there is also cooperation between different subsets of T cells. The HLA system plays a role in all these cellular interactions, some of which are illustrated in figure 7. The foreign antigen is taken up by the macrophages, which 'present' it to a subset of T lymphocytes called *T helper lymphocytes*. These cells become activated and are now able to help another subset of T lymphocytes to develop into *effector T lymphocytes* which can kill cells carrying the foreign antigen, e.g. virus-infected cells. T helper lymphocytes also interact with B lymphocytes and help them to transform into plasma cells which produce large amounts of IgG antibody directed against the foreign antigen.

The MHC genes responsible for the monocyte/T helper lymphocyte and T helper lymphocyte/B lymphocyte interactions are probably the same and map within the I region of H-2 in mice and seem to be close to the HLA-D/DR locus in man, although in man this has as yet only been established for the genes controlling monocyte/T helper cell cooperation. In fact, there is increasing belief that the gene products responsible for the interaction are the HLA-D/DR antigens (in man) and the Ia antigens (in mice) themselves. These antigens are present both on macrophages and on B lymphocytes, but absent on T helper lymphocytes. In order that a T helper cell can be activated, it is necessary that it recognizes not only foreign antigen, but also the HLA-D/DR antigen on the monocyte presenting the foreign antigen to the T helper cell. The cooperation between T helper cells and B lymphocytes also requires the involvement of HLA-D/DR homologues in animals, but this has not yet been shown in man. It would appear that the B lymphocytes, in order to receive help from T helper cells, must carry the same D/DR antigens as the macrophages presenting the foreign antigen to the T helper cells.

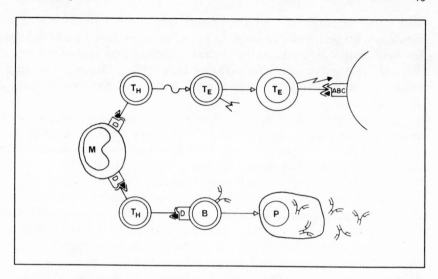

Fig. 7. Simplified theoretical model of thymus-dependent immunity. M = Macrophage, T_H = T helper lymphocyte, T_E = T effector lymphocyte, B = B lymphocyte, and P = plasma cell. The foreign antigen, symbolized by the black triangle, is taken up by the macrophage and presented together with HLA-D antigen to T helper cells which, when thus activated, can trigger precursor cells of T effector cells to become active killer cells. These cells react both with foreign antigen and with HLA-ABC antigen on the target cell. The same or another set of T helper cells helps B lymphocytes to transform into IgG antibody-producing plasma cells. This figure shows only one of the several possible models for the cellular cooperation (combined presentation and a single T-cell receptor), but it could be replaced by any of the five other models illustrated in figure 8.

While the HLA-D/DR antigens are involved early in the immune response, i.e. the triggering of T helper cells, and also later in the induction of IgG antibody production against thymus-dependent antigens, the HLA-ABC antigens play a role mainly in the later steps, i.e. the lysis of cells by T killer cells. In fact, much of the above scheme has been deduced by homology from knowledge of the involvement of H-2D and H-2K antigens (in mice) and of HLA-ABC antigens (in man) in the lysis of virus-infected or hapten-conjugated cells by T killer lymphocytes. The situation may be illustrated by the findings of *Doherty and Zinkernagel* (80), and *Zinkernagel and Doherty* (347). It has long been known that mice infected with lymphocytic choriomeningitis (LCM) virus develop T killer lymphocytes which can be recovered from the spleen and which specifically can lyse LCM virus-infected cells. *Doherty and Zinkernagel* observed, however, that these killer cells do not lyse LCM virus-infected cells from all strains of mice but only from strains which share H-2D and/or H-2K antigens with the

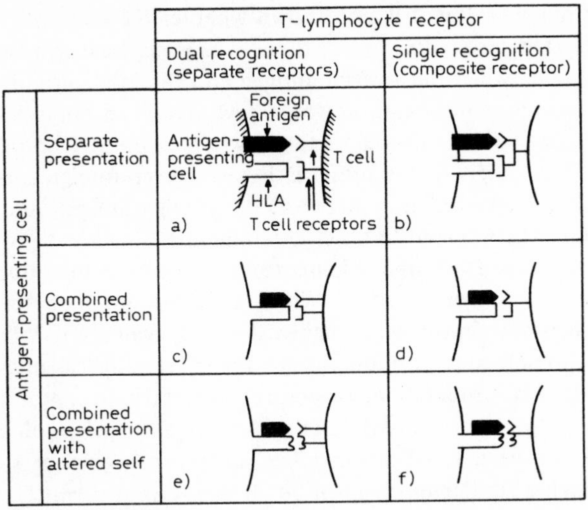

Fig. 8. Models for molecular interactions in thymus-dependent immunity. There may be one (b, d, and f) or two (a, c, and e) receptors on the T lymphocyte, and on the antigen-presenting cell, the foreign antigen may be separate (a and b) or combined (c, d, e, and f) with an HLA molecule; in the latter case, an alteration of the HLA molecule may take place (e and f). When the T lymphocyte is a T helper cell, the antigen-presenting cell is a macrophage and the HLA molecule is probably HLA-D/DR antigen. If it is a T effector lymphocyte, the antigen-presenting cell may, for example, be a virus-infected cell and the HLA molecule is an HLA-A, B, or C antigen.

mouse in which the T killer cells were developed. The Ia antigens are not involved in this process. Therefore, the T killer cells react not just with the virus on the infected cells, but both with the virus and the H-2D and H-2K antigens on these cells. It was on the basis of these observations that they (80, 347) advanced the 'dual recognition' and 'altered self' hypotheses which are illustrated in figure 8. Unfortunately, these two hypotheses are not mutually exclusive, which has given rise to some confusion. The term 'dual recognition' means that there are two separate receptors on the T lymphocytes, one for foreign antigen and one for autologous MHC products. 'Altered self' means that an interaction takes place between the foreign antigen and the MHC product and that this creates an alteration of the MHC product. It appears from figure 8 that it is possible to have a combination of 'altered self' and recognition by two different T-cell receptors ('dual recognition') simultaneously. The essential point to note is that the T lymphocyte must have a receptor both for foreign antigen and for autologous MHC determinants, but it is not known whether these receptors are separate or combined. It is also uncertain if there is any physical interaction (combined presentation) between the foreign antigen and the MHC product on the presenting cell. A combined presentation would have several advantages from a theoretical point of view because it could explain why certain MHC products are apparently better than others in the presentation of a given foreign antigen. This could be due to a higher affinity between the foreign antigen and the 'superior' MHC determinant compared to other determinants.

The cell-cell cooperation illustrated schematically in figure 8 may operate for the different kinds of T lymphocytes. When T helper lymphocytes are involved, the MHC-specific receptor is probably directed against HLA-D/DR determinants or their homologues in animals, and the presenting cell may be either a macrophage or a B lymphocyte. In contrast, the MHC receptor on T effector lymphocyte has specificity for HLA-ABC antigens or their homologues, and the presenting cell may be any cell carrying ABC antigens. Suppressor T cells probably operate through a similar mechanism but it is not known which MHC products are involved.

According to this scheme, it is possible — perhaps even likely — that the D/DR antigens themselves may be the Ir determinants responsible for the variation in immune responsiveness.

The nature of the T-cell receptors is unknown (58). It has been suggested that it consists of the variable part of one heavy chain of immunoglobulin, because T and B lymphocytes have similar idiotypes (126). If there are two T-cell receptors (dual recognition), these may be of the same or different structure, but must be of different specificity.

There is some evidence that a direct contact between the collaborating cells in the cell-cell interaction is not always necessary, but that it may be substituted by soluble factors carrying MHC products (203, 301).

The involvement of HLA-D antigens in the cooperation between monocytes and T lymphocytes was shown independently by *Bergholtz and Thorsby* (30) and by *Hansen et al.* (111, 112). Both of these groups used PPD (purified protein derivate = tuberculin) as foreign antigen and purified T lymphocytes from BCG vaccinated individuals. They found that the T lymphocytes will only transform to PPD *in vitro* when monocytes are present in the culture. Moreover, the monocytes must share HLA-D antigens with the T lymphocytes in order that an optimal reaction to PPD takes place. By analogy with the animal experiments, these observations were interpreted as evidence that the T lymphocytes must recognize both PPD and 'autologous' HLA-D antigens before they are stimulated.

It has also been demonstrated that the HLA-ABC antigens in man play a similar role as the H-2D and H-2K antigens in mice: (i) *Goulmy et al.* (103) showed that this is true for the lysis of male cells (carrying the H-Y antigen) by female lymphocytes sensitized against male cells; (ii) *Dickmeiss et al.* (79)

demonstrated this role in the lysis of lymphocytes coated with the hapten dinitrofluorobenzene (DNFB), and (iii) *McMichael et al.* (188) showed that lysis of influenza virus-infected cells is also restricted by HLA-ABC antigens. Interestingly, one antigen, HLA-A2, seemed to be superior in the former two systems, while another, HLA-B7, was superior in the latter. This 'preferential' restriction has also been observed in mice (42, 271) and all in all, this phenomenon favors the 'presentation' hypothesis according to which certain HLA factors are superior in the presentation of certain foreign antigens while other HLA factors preferentially present other foreign antigens. If the same phenomenon holds for the action of the Ia antigens, this would easily explain why some antigens are associated with high and others with low responsiveness.

At the time of writing, one of the most fascinating questions concerns the mechanism which causes the MHC restriction of the T-lymphocyte receptors. It has been known for some time that T lymphocytes mature in the thymus. For example, individuals born without a thymus do not develop functional T lymphocytes and thus, no thymus-dependent immunity. Very recently, an elegant but complex series of experiments by *Zinkernagel et al.* (346) provided evidence that the thymus is also responsible for 'teaching' immature precursors of T lymphocytes to be restricted by MHC. In these experiments, X-irradiated and thymectomized mice received T-lymphocyte-depleted bone marrow grafts and X-irradiated thymus grafts. By varying the H-2 types of the recipient, the bone marrow donor, and the thymus donor, these investigators showed that the T lymphocytes maturing in the recipient mice were restricted by the MHC antigens of the thymus donor. The exact mechanism by which this MHC restriction is brought about in the thymus is unknown, but it is possible that MHC antigens on the thymic epithelial cells select T-lymphocyte precursors which have receptors for autologous MHC determinants and that only such cells are allowed to develop into mature T cells. Indeed, it has been suggested that the contact with MHC antigens in the thymus is not only responsible for the MHC restriction but also for the development of the large number of different clones of T lymphocytes, each with their own specificity for a foreign antigen (39b). If this is true, we may be closer to solving the problem of the 'origin of antibody diversity', i.e. why can every individual form hundreds of thousands of different antibodies, each with their own specificity?

The MHC restriction phenomenon may explain several enigmatic observations. For example, it has long been a mystery why as many as between 2 and 5% of lymphocytes from an unimmunized individual will respond in an MLC to allogeneic cells from just one individual. Now, the MLC reaction may be considered a reaction against 'altered self' HLA-D antigens, i.e. the responding T lymphocytes recognize the allogeneic HLA-D antigens as altered self. This also provides an explanation for the HLA system being such a strong immunological barrier in transplantation.

In conclusion, the MHC serves a major biological role by being involved in many of the cooperations between various cell types belonging to the immune system. Many of these cells are not fixed in a given tissue, but must be able to enter almost all parts of the body to conduct their defensive services. Under these conditions, there must be mechanisms which ensure that the moving cells can find each other and cooperate in the adequate sequence. The various MHC determinants may — together with foreign antigen — be considered specific signals which help in selecting the correct T-lymphocyte subgroup to be activated.

8.2.2. Other Hypotheses on the Biological Function

(1) *Burnet* (47) believes that a major function of these histocompatibility antigens is to prevent infectivity of tumor cells. If histoincompatible tumor cells are transferred from one individual to another, the immune system of the recipient is likely to recognize the foreign HLA antigens and to raise an immune attack against these cells. *Burnet* takes the extreme rareness of contagious tumors as evidence that histocompatibility antigens are very effective as a protective mechanism.

(2) A somewhat related theory was discussed earlier by *Nandi* (206) who suggested that the antigenicity of histocompatibility antigens serves a function by being incorporated in the outer membrane of virus particles which are thus made more easily recognizable by the host infected with virus from a histo-incompatible individual. *Simonsen* (270) has given indirect evidence that this mechanism may operate between species.

(3) The alloantigenic properties may also be of importance by creating differences between a maternal and a fetal organism which could be necessary to ensure the integrity of these. However, experience with inbred animals, which are identical at all loci, does not seem to support this assumption.

In general, it may be worth noting that self-recognition systems are known also in lower organisms without immunity systems (e.g. sponges), and such systems may thus be important in most multicellular organisms. In fact, it seems probable that both histocompatibility and immune systems of higher species originate from the self-recognition systems now seen in primitive organisms.

One advantage of these three theories is that they explain the large polymorphism of the HLA system: the larger the polymorphism, the more effective would these mechanisms be.

(4) Some histocompatibility antigens are *differentiation antigens* in the sense that they are present only on some cell types. For example, DR antigens are present on B lymphocytes, macrophages, and some other cells, but not (or

only to a small extent) on T lymphocytes or platelets. On the other hand, the A, B, and C antigens seem to be present on all cells except erythrocytes. The term 'differentiation antigen' (37) is often extended to mean that these antigens are necessary for the differentiation process itself or for the specific functions of the differentiated cells, and this may be true of the DR antigens which apparently are involved in the cooperation between B and T cells.

The T-t complex in mice (p. 38) is a differentiation antigen system involved in embryogenesis, and if the homology between the T-t antigens and H-2 is confirmed, the HLA factors may be considered differentiation antigens for some cellular interactions in the adult.

There is one important difficulty to solve if the main function of HLA factors is differentiation: why should they then need to be polymorphic?

(5) *Jerne* (128) suggested that HLA antigens are of fundamental importance for the development during fetal life of a large number of different clones of immunocompetent cells each with its own antibody specificity. Very recently, this hypothesis has gained significant support by the observation that immature immunocompetent T lymphocytes, when matured by contact with epithelial cells in the thymus, apparently learn to recognize altered or unaltered self HLA antigens (346). *Jerne* (39b) suggests that the T lymphocytes have two receptors and that mutants of one of these are selected for in the thymus, while the other receptor maintains its anti-self specificity.

The above list of theories is not complete, but merely illustrates the present way of thinking.

8.3. Maintenance of Polymorphism

Concerning the high degree of polymorphism of HLA, the genetic question remains to be solved whether it is maintained by natural selection, by genetic drift (or related 'neutral' mechanisms), or both. By comparing genetic distances between various populations for a variety of blood groups, *Bodmer et al.* (38) observed that the HLA system — in spite of its enormous diversity — was among those which showed the least variation and took among others this as evidence that natural selection is indeed involved in maintaining this polymorphism. The associations observed between HLA and various diseases meet *Ford*'s (97) expectation that blood group antigens are not selectively neutral. At the same time it should be stressed that most of the diseases in question are quite rare (juvenile diabetes), occur after the onset of fertile age (ankylosing spondylitis), or do not influence fertility (i.e. 'fitness') appreciably (psoriasis), and thus it is extremely unlikely that these associations alone can explain the entire HLA polymorphism. However, they do show that HLA factors may be subject to selection, and a major step in the same direction would be obtained if it could be shown that some of the diseases involving large selective pressures (e.g. present or past epidemics) are associated with HLA.

Although perhaps contradictory at first thought, it seems to us that selection *against* individual HLA factors cannot be very strong, or else they would have been eliminated. Conceivably there may be quite a freedom for mutations to occur within the HLA system, but the mutants must then present an advantage to become fixed in the population (95). As discussed by *Bodmer* (37), selection for rare alleles could explain the polymorphism.

Advantage of heterozygous individuals over homozygous ones is a mechanism which can explain a stable polymorphism (49). There is now good theoretical reason to believe that individuals who are heterozygous at the MHC may be at an advantage. For example, heterozygosity at an Ir locus might ensure a high immune response to two different sets of foreign antigens corresponding to the two Ir determinants. Moreover, individuals who are heterozygous at the HLA-ABC loci have twice as many antigens which can cooperate in the lysis of virus-infected cells as have homozygotes. In fact, *Doherty and Zinkernagel* (80) observed already in their original experiments that H-2 heterozygous mice generated more effective killing of virus-infected cells than did homozygous animals.

However, there is as yet no definite evidence, e.g. from Hardy-Weinberg or segregation analyses or from studies of old people (115), that HLA heterozygotes are at an advantage. Nevertheless, it should be noted that such studies have mainly been done in developed countries, where natural selection is probably not as active as it may still be in developing areas.

Finally, for the sake of completeness, we should mention that drift or related mechanisms obviously must have played a role for the HLA gene frequencies as they are found in various populations today. It is more doubtful, however, whether drift alone as suggested by *Degos and Dausset* (74) can explain the existence of gametic association (linkage disequilibrium) between the various segregant series. The observation of disease resistance confined to one or a few haplotypes would be of considerable help in our understanding of this extraordinary linkage disequilibrium. It may be speculated that the combination of some specific *Ir (= D/DR?)* gene with a specific *HLA-A, B,* or *C* gene in one haplotype could be particularly valuable in the development of specific killer cells where both Ir determinants (in the generation of helper cells) and ABC antigens (in the killing phase) are necessary. However, experiments exploring this possibility have not been reported.

9. HLA in Clinical Medicine

The main reason for the intensity of the research leading to a rapid — though far from complete — unravelling of the HLA system obviously was the hope to improve the results of clinical organ transplantation primarily of cadaver kidneys. As discussed below, this hope has as yet been only moderately fulfilled. However, it now appears that the HLA system has important implications in much larger areas of clinical medicine than that concerned with organ grafting. For example, immunization with HLA antigens through pregnancy and/or blood transfusions creates problems both in terms of febrile reactions to transfusion, and in particular in the treatment of thrombocytopenic patients with platelets. Of much greater importance, however, is the fact that the HLA system is involved in the development of a variety of diseases, which has led to new ways of thinking in relation to the genetics, pathogenesis, and etiology of several of these diseases.

9.1. Transplantation

The rejection of a transplant is due to the immunological reactivity of the recipient towards antigens present on donor cells and lacking in the recipient. The observation that kidneys from HLA identical and ABO blood group compatible siblings have almost perfect survival in the recipient (e.g. 67, 215, 279) shows that HLA and ABO are the histocompatibility systems of key importance in man. A similar high degree of graft survival is not seen for kidneys from unrelated donors even if they are also ABO compatible and as HLA identical as possible in terms of the A, B, and probably C series antigens. Thus, these antigens cannot be the only HLA factors determining graft survival. Before discussing these problems further, we find it useful to recall a few definitions often used in the field of transplantation and to give a brief summary of transplantation immunology.

Transplantations are usually classified according to the relationship between the donor and the recipient: (1) in *xeno*(hetero)transplantations the donor and recipient belong to different species; (2) *allo*transplantations take place between genetically different individuals belonging to the same species; (3) in *iso*transplantations the donor-recipient pairs are genetically identical (monozygotic twins, animals of the same inbred strain, or F_1 hybrids between two such strains), and (4) *auto*transplantation is the removal of tissue from one part of the body to another within the same individual. It is worth noting that the term 'allo' for transplantations is used in the same sense as 'iso' is in blood-grouping serology. From an immunological point of view, iso- (and auto-)transplantations are usually without interest because the recipient cannot recognize the transplant as foreign, and accordingly, the survival of the transplant is as long as it would have been in the donor. In the following we shall deal almost exclusively with allotransplantations. Concerning heterotransplantations it suffices to state that these usually meet such strong immunological barriers that rejection (cf. below) invariably occurs despite all measures to combat it. The discussion of pregnancies and blood transfusions, which are special cases of allotransplantations will each be given a section of their own.

9.1.1. Transplantation Immunology

When allogeneic tissue is transferred from one individual to another, some of the immunocompetent lymphocytes of the recipient will recognize the foreign transplantation antigens of the donor. There are two kinds of immunocompetent lymphocytes (205, 240): *T lymphocytes* (T cells) which have been matured in the thymus, and *B lymphocytes* (B cells) named after bursa fabricii in birds because they are matured there, while the homologous organ is unknown in other vertebrates including man. Both of these cell types carry receptors for foreign antigens on their surface, each cell carrying only one specificity. The B-cell receptors are immunoglobulins, while the nature of the T-cell receptor is unknown but it may be related to immunoglobulins (126). When the receptors react with the corresponding foreign antigen (e.g. on transplanted cells), the lymphocytes in question are stimulated to transform into blast cells which divide (as it also happens in the MLC test; cf. p. 86).

When T cells are stimulated, at least two kinds of progenitor cells are formed (fig. 9): *killer cells* which react specifically with the transplanted cells and destroy them either by direct contact or by the release of so-called lymphokines, which in turn also cause vascular permeability and increased phagocytosis among others. The other T-cell progenitors are *T-memory cells* which probably are principally identical to the original lymphocytes, i.e. they have the same specificity but they are now present in much higher numbers which is the explanation for the usually more accelerated and vigorous reaction seen when a subsequent transplant of the same donor type origin is performed.

Stimulation of B lymphocytes with a foreign transplant also gives rise to two kinds of progenitor cells: *plasma cells* secreting antibodies specifically reactive with the antigen(s) in question, and *B-memory cells* probably similar to those originally stimulated. As mentioned earlier, B cells often need the cooperation of T cells to become antibody-secreting cells.

In first transplants, the T-cell activation is the all important factor responsible for the rejection of the graft. In subsequent transplantations, the presence of preformed antibodies in the recipient may cause hyperacute rejection of the graft within minutes when they

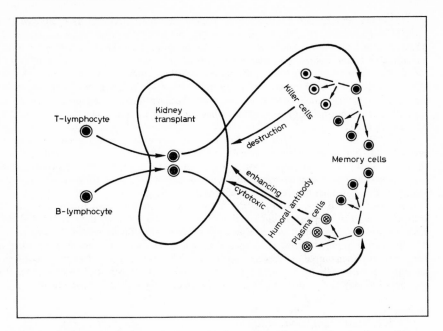

Fig. 9. Simplified diagram of the activation of the immune system by a kidney allograft. For further explanation, see text.

combine with antigen from the transplant and activate complement which damages the cell membranes and causes platelet aggregation leading to vascular thrombosis.

The ordinary rejection process can be circumvented or decreased either by minimizing the antigenic disparity between donor and recipient and/or by decreasing the immunologic capacity of the recipient. The former is obtained by selecting donor-recipient pairs as antigenically identical (compatible) as possible, and the latter by various immunosuppressive regimens. Many methods of immunosuppression exist, but for practical purposes, certain drugs, such as azathioprine and steroids, and heterologous anti-lymphocyte serum (ALS) or globulin (ALG) directed against human lymphocytes have proved most useful.

It is kind of a mystery why definitely 'incompatible' grafts can in some cases survive for many years. Two main mechanisms have been held responsible for this phenomenon: tolerance and enhancement. *Immunological tolerance* means decreased or abolished immune responsiveness towards one or more specific antigens. The classical example of tolerance is the inability of an individual to react immunologically to its own 'self' antigens (46). This natural self-tolerance is achieved in fetal life, but tolerance against foreign antigens can be induced artificially by injecting newborn and even adult animals with very high doses (high zone tolerance or immune paralysis) or repeatedly with very small doses (low zone tolerance) of a given antigen. Tolerance was originally thought to indicate the absence of lymphocytes capable of reacting specifically with the antigen, and this may be true of self-tolerance. However, recent research seems to indicate that reactive lymphocytes are present, but that their reaction with antigen does not lead to the usual formation of plasma cells or killer cells. Tolerance is obtained more easily in the immunosuppressed individual.

In animal experiments, it has proven possible to obtain impressively prolonged graft survival by pretreatment with so-called *anti-idiotypic antibodies* (33). Such antibodies are directed against the immunoglobulin receptor site for foreign antigen. It is possible to render an animal tolerant against transplantation antigens of type 'A' by special immunization procedures leading to production of antibodies against antibodies to A (i.e. anti-anti-A), and this tolerance may be brought about by the anti-anti-A reacting with and destroying lymphocytes having anti-A specificity. Anti-idiotypic antibodies may play a role in the regulation of the normal immune response where they may interact in a large and complicated network (129).

Another mechanism which may bring about tolerance in the presence of antigen-reactive cells is the formation of specific *suppressor cells* which have attracted increasing interest recently (29, 323). Suppressor cells are a subset of T lymphocytes and HLA-specific suppressor cells have been described in parous women (187, 311). However, their role in kidney-graft survival has not yet been established.

Enhancement is due to circulating antibodies which react specifically with the incompatible antigen in question but which protect rather than destroy the transplanted cells. Like tolerance, enhancement is a complex phenomenon which is far from fully understood. The protection brought about the enhancement may be due to the possible 'masking' of antigens with noncomplement-fixing antibodies which would prevent complement-fixing antibody or killer cells from reacting with their target cells. Recent evidence indicates that enhancement may also be due to circulating antigen or antigen-antibody complexes.

9.1.2. Transplantation Antigens

Histocompatibility antigens are controlled by a number of nonlinked genetic systems. For example, the number of such systems have been estimated to about 20 for skin grafts in mice (152). In man, the observation that skin even from HLA-identical and ABO-compatible siblings never survives indefinitely indicates that this species also has a considerable number of transplantation antigen systems. The same has been found true of all other vertebrates studied so far. In all of these species, however, one system has proven strikingly more important than the other systems within the species in question. The strong or *major* histocompatibility system in man is the HLA system. The proof of this has been most precisely pointed out by *Ceppellini et al.* (53), who observed the skin graft survival times in the various HLA combinations listed in table XIII. It appears that skin from HLA-identical siblings survives on an average for about 3 weeks, HLA haploidentical skin survives for about 2 weeks, while skin from two-haplotype different siblings is rejected about 12 days after transplantation. The important point is that this latter survival time is identical to that seen in random transplantations between unrelated donor-recipient pairs: If there were one or more other strong histocompatibility systems in man, the two-haplotype different siblings pairs would be identical for at least a quarter of these, and this should have been reflected in a prolonged survival in some of these related combinations. Additional evidence that HLA is the strong system in man appears from the fact that kidneys from HLA-identical siblings have a substantially

Table XIII. Influence of HLA in transplantation

	Donor	Recipient	Skin graft (mean survival) days	Kidney graft (1-year graft survival), %	Bone marrow graft
Siblings	$a/c \rightarrow$	a/c	20.0	90	often successful
	$a/d \rightarrow$	a/c	13.8	70	failure
	$b/d \rightarrow$	a/c	12.5	60	failure
Unrelated	$x/y \rightarrow$	a/c	12.1	50	failure

Letter *a*, *b*, *c*, *d*, *x*, and *y* indicate different HLA haplotypes. All combinations are ABO compatible. Skin graft data are from *Ceppellini et al.* (53), and kidney data are approximate values based on *Opelz et al.* (215) and *Solheim et al.* (279).

better survival than those from other relatives, and from the almost inevitable fatal outcome of engraftment with HLA nonidentical bone marrow (cf. below).

In addition to the pronounced differences in survival times for skin grafts incompatible for major and minor systems, we may mention that major antigens are particularly important for organ graft survival and that preimmunization against minor antigens usually has a large effect on the reduction of the survival time of secondary skin grafts incompatible for the same minor antigen.

Minor Transplantation Systems in Man. Theoretically, all alloantigen systems are histocompatibility systems. Thus, all blood group systems may be considered histocompatibility systems. For practical purposes, however, only the ABO blood group system is of fundamental clinical importance. This is due to the fact that all normal individuals of groups O, A, and B have preformed antibodies against the ABO antigens which they lack. Like HLA antigens, A and B blood groups antigens are present on most human cell types, and accordingly, this system must be taken into account in all transplantations, except when the recipient suffers from severe combined immunodeficiency. Recently, evidence has been provided that the Lewis blood group system may also play a role in kidney transplantation (217). Some transplantation antigens are *tissue-specific*, i.e. they are present only on cells from certain tissues. This is true of most red blood group cell antigens such as Rhesus and others, and these need only to be taken into account in blood transfusions, and in bone marrow transplantation, perhaps only when the recipient is preimmunized. Apart from blood group antigens, our knowledge of tissue-specific alloantigens in man is poor, but there is some evidence in favor of the existence of kidney-specific transplantation antigens (222). The fact that kidneys from HLA-identical siblings are quite often subject to rejection episodes (which are usually successfully circumvented by immunosuppressive therapy) reflects the existence of weak histocompatibility systems in man. Extrapolation from animals indicates that most weak histo-compatibility systems have a limited degree of genetic polymorphism, i.e. there are only few alleles within each system.

9.1.3. Kidney Transplantation

As mentioned earlier, when the donor is a monozygotic twin, a transplant is permanently accepted, and this is, of course, also true for kidney transplanta-tion: recipients of such kidney-*iso*-grafts need not receive immunosuppression at all, but it seems worth noting that the original disease present in the recipient (e.g. glomerulonephritis) may attack the transplant.

From an immunogenetic point of view, the next best kidney donor is an HLA-identical and ABO-compatible sibling — such kidneys have a very good survival rate (table XIII), but even in this situation some immunosuppression is almost always needed to avoid rejection due to incompatibility for minor histocompatibility antigens. Kidneys from HLA haploidentical relatives do not survive nearly as well as those from HLA-identical siblings (table XIII). ·

Many uremic patients have no relatives who are suitable as donors, and as there are also ethical problems involved in removing kidneys from living donors, most patients are transplanted with cadaver kidneys removed shortly after the death of the donor.

In order to obtain the best HLA matching between donor and recipient in cadaver transplantation, cooperation between several centers is necessary. All suitable patients with chronic uremia within a region are HLA typed and listed in a registry. The vast majority of the patients have reached a state where dialysis is necessary. When a donor becomes available in a center, HLA typing is performed and the two recipients for whom these kidneys are most HLA compatible are selected from the file. When the kidneys have been removed, perfused, and cooled, they are transported to the recipient centers where the transplantation is performed. Cold ischemia times are usually kept below 24 h.

Until recently, HLA typing has included only the classical antigens of the A, B, and occasionally, the C segregant series (the latter are, however, rarely taken into account in the matching) because HLA-D typing cannot yet be done within the time available. Distinction is usually made between related and unrelated donors because HLA identity between siblings almost always means identity for all genes within the HLA system whereas this is rarely so for HLA 'identity' (A and B series only) between unrelated individuals. The nomenclature with letters (table XIV) is somewhat inconsequent as the worst possible match F = positive cross-match, is listed before G = incompatibility for four antigens. These letters should not be confused with the designations for the A, B, C, and D antigens. A more logical grading of the matches is to indicate the number of antigens

Table XIV. Match grades

Match grade	Interpretation	New grading
A1	HLA identity between siblings	0
A2	HLA identity between parent-child	0
A3	HLA identity between unrelated	0
B	recipient has an antigen lacked by the donor (minor incompatibility)	0 (1)
C	incompatibility for one antigen	1
D	incompatibility for two antigens	2
E	incompatibility for three antigens	3
F	positive cross-match	
G	incompatibility for four antigens	4

Usually only HLA-A and B antigens are taken into account. The term 'B match' is used because there is possibility for one or more incompatibilities; e.g. an HLA-A1; B7,8 donor (to an A1,3;B7,8 recipient) may be A1/1 homozygous or A1/X heterozygous with X as an unknown A antigen. The new grading simply indicates the number of antigens mismatched. Note that some workers conversely give the number of antigens shared by the donor and recipient.

mismatched. Conversely, some investigators give the number of antigens shared by the donor and recipient. If the lymphocytotoxic *cross-match* between recipient serum and donor lymphocytes is positive, the kidney is usually rejected in a hyperacute way within minutes to few hours after the blood supply of the kidney has been established in the recipient. Accordingly, a cross-match should be carried out before the transplantation if this is at all possible, and it must be done when the recipient is known to have HLA antibodies, which is the case for about a quarter of the recipients. There are some indications that cross-matching with the LALI test (cf. p. 83) may be superior to the ordinary lymphocytotoxic test (320). Recipient antibodies against donor DR antigens do not seem to influence graft survival (92, 199).

It came as a surprise for most HLA workers that matching for as many HLA antigens as possible did not increase the graft survival of unrelated kidneys to any great extent. Indeed, it has been claimed that there is no correlation between the match grade and graft survival (215) and in a few centers all ABO compatible kidneys are accepted when the cross-match is negative. However, in a number of studies each comprising a large number of patients, significant correlation between the number of HLA antigens mismatched and graft failure has been observed (68, 154, 261, 279). There is some evidence that matching for B series antigens is more important than matching for A series antigens. This would be in line with the fact that identity for B antigens in some cases would ensure identity for D antigens too, because of the association between B and D antigens caused by the linkage disequilibrium between the corresponding genes. It is to be expected that considerable improvement of the graft survival can be achieved when rapid typing for D antigens and/or DR antigens closely associated with D makes it possible to match cadaver kidneys for D antigens. Indeed, preliminary reports of retrospective analyses of cadaver kidney survival indicate that compatibility for DR antigens (7, 179, 223, 319) and a low MLC reaction of the recipient against donor cells (257, 310) are associated with a strikingly good survival.

Although certainly exceptionally, a few cases of perfect graft survival have been observed in spite of a positive cross-match. This is believed to be due to enhancement, but unfortunately attempts to characterize enhancing antibodies and to separate them from damaging antibodies have failed so far. It has also been tried to induce passive enhancement by means of noncomplement fixing $(Fab)_2$ fragments of HLA antibodies directed against donor antigens, but it is too early to judge the results of these experiments in humans (22).

Within the last few years it has become clear that there is a striking association between kidney graft survival and prior blood transfusion given to the recipient — see *Opelz and Terasaki* (214) for review. The effect of transfusion seems to be related to the number of units of blood given. There are at least two possible explanations for these observations. Firstly, transfusion therapy may serve as a selection procedure because good antibody responders are likely to form multispecific antibodies, making it very difficult to find cross-match negative kidneys, while poor responders still have a high chance of being transplanted. Secondly, transfusion therapy may in some patients induce a state of unresponsiveness, e.g. due to enhancement or suppressor cells. The first of these mechanisms probably explains a considerable part of the association observed, and at the time of writing the following crucial questions cannot be answered: Will a liberal transfusion policy result in a high number of hyperimmunized dialysis patients for whom a cross-match negative kidney will never be found? Should such responders in fact never be transplanted or is it possible that D/DR-matched kidneys will survive even in such potentially good responders who have not yet been transfused? In any case, there is a great need for a reliable *in vitro* test to assess the responsiveness of a potential kidney recipient without giving transfusion.

9.1.4. Bone Marrow Transplantation

In most transplantations, the immunological problems have been only in one direction: host-versus-graft reaction. In addition to this reaction, bone marrow grafts involve a reaction in the opposite direction: a *graft-versus-host (gvh)* reaction because bone marrow inevitably contains immunocompetent cells —

or precursors thereof — which can react against the antigens of the recipient (25).

Three groups of patients have so far been treated with allogeneic bone marrow with some success: (1) patients with severe combined immunodeficiency who apparently lack the bone marrow stem cell which can develop into mature T and B lymphocytes; (2) patients suffering from aplastic anemia, or (3) leukemia. Theoretically, however, other and more common diseases such as the hemoglobinopathies may be candidates for bone marrow transplantation, but as discussed below, this procedure is met with such great immunological problems that it is still at an experimental stage.

Technically, bone marrow transplantation is very simple: the bone marrow is usually aspirated in 1 ml aliquots from multiple punctures of the iliac crest, the spine, and sometimes the sternum. In immunodeficiency only about six punctures are needed and these can be done in local anesthesia. In aplastic anemia and leukemia about 50—100 punctures may be needed because about 3 $\times$ 10^8 nucleated cells per kilogram body weight are needed. The bone marrow is usually given simply by the intravenous route after simple filtration to remove osseous fragments.

About 30 infants with severe combined immunodeficiency have so far been successfully treated with bone marrow grafts. Untreated, these patients succumb to infections within the first year of life. The Seattle group has obtained a 1-year survival rate of about 50% in aplastic anemia and 20% in leukemia (286, 305). However, continuation of the leukemic process in the engrafted cells has been reported (304).

Obviously, patients with severe combined immunodeficiency cannot raise a host-versus-graft reaction, but in immunocompetent recipients, heavy immunosuppression is usually necessary to obtain a 'take' of the graft. A fatal gvh reaction may be seen after transfusion of blood containing viable lymphocytes in severe combined immunodeficiency, but can be avoided by X-irradiation or freezing and thawing of the blood product. The gvh reaction usually starts one to a few weeks after the allogeneic cells have been injected, and it is characterized by dermatitis, liver and spleen enlargement, diarrhea, thrombocytopenic purpura, and infections due to neutropenia. If the donor is not HLA-D compatible with the recipient, this reaction is invariably fatal. As in kidney transplantation, the bone marrow donors of choice are HLA-identical siblings, but successful grafting has been reported with HLA-D-identical donors differing for some of the antigens of the A and B series (81, 101a, 121, 171, 216). In severe combined immunodeficiency, the ABO system needs not be taken into account, and it is even possible in aplastic anemia and leukemia to have successful engraftment of ABO-incompatible bone marrow provided special measures are taken to reduce the titers of anti-A and/or anti-B in the recipient (44), e.g. by plasmapheresis and infusion of blood of donor ABO type.

In contrast to the situation in kidney transplantation where transfusion therapy may have a favorable influence on graft survival, the chances of successful engraftment with bone marrow from HLA-identical siblings decrease with the number of transfusions given to patients with aplastic anemia prior to transplantation. This indicates that preimmunization against minor transplantation antigens play a considerable role. In general, young patients who have received no or few transfusions have the best prognosis (285). Moreover, there is some evidence that transplantation between the two sexes are more difficult than when donor and recipient are of the same sex.

9.1.5. Transplantation of Other Tissues

Except in patients with severe deficiencies of the thymus-dependent part of the immune system, *skin allografts* are almost invariably rejected even in highly immunosuppressed recipients, and thus skin allotransplantation from cadaver donors is used only as a temporary cover, e.g. of extended burns.

Heart and liver transplantations are performed in a few centers, and 1-year survival rates of about 45% have been reported for heart transplants (105). For both organs, great surgical skill is needed. The series reported so far are too small to give indications as to the importance of HLA matching in these transplantations.

There is some evidence that HLA matching may improve the results of *corneal* transplantation (21, 89, 145). As it is possible to store cadaver corneae, it should be possible to establish a 'bank' of HLA-typed corneae provided blood or lymph nodes from the cadaver can be obtained rapidly enough to allow HLA typing after death.

Lung and pancreas transplantations have met with such great technical problems that they are still at a very early experimental stage.

Bone and cartilage may lack a significant amount of HLA antigens, and as these tissues serve mostly as a matrix for the ingrowth of recipient cells, the importance of HLA matching has not been evaluated.

Transplantation of *fetal thymus* or *cultured thymic epithelium* has been used in some cases of severe immunodeficiencies, but it is too early to judge the value of this treatment (122a).

9.2. Transfusion

In modern blood transfusion therapy attempts are made to utilize the donor blood as rationally as possible by providing the patients only with the components of the blood which they need. Concerning the formed elements of the blood, for example, patients with chronic anemia should be given only red cells without the other cells and plasma, whereas bleeding can be prevented in thrombocytopenic patients by blood platelets. In practice, however, it is quite laborious to produce red cell concentrates devoid of leukocytes and platelets, and accordingly most anemic patients receive packed red blood cells which also contain most of the leukocytes and platelets of the blood. These cells contain HLA antigens and HLA immunization is a frequent consequence of repeated transfusions. We discuss below the significance of HLA immunization in relation to febrile transfusion reactions, platelet and granulocyte transfusion.

9.2.1. Febrile Transfusion Reactions

In about 2% of all blood transfusions, the patients react with shivering and a rise in temperature during or within a few hours after the infusion (196). In most cases, these reactions are due to recipient HLA antibodies directed against donor HLA antigens. This causes destruction of donor granulocytes with liberation of pyrogens. In other cases, such reactions are due to other pyrogens, or to red cell incompatibility, to granulocyte-specific antibodies, while lymphocyte and platelet destruction probably do not cause such reactions. When caused by HLA incompatibility, the reactions are generally considered harmless though unpleasant for the recipient. However, it cannot be ruled out that they may be dangerous to patients who are severely ill, and it has been suggested that they may run a fatal course in rare cases (127). For these reasons and to avoid the interruption of the infusion, which is recommended in case of untoward transfusion reactions, it is advisable to transfuse patients known to have leukocyte antibodies with granulocyte or leukocyte poor blood.

Finally, for the sake of completeness, it may be noted that the infusion of plasma or whole blood containing antibodies directed against recipient HLA antigens may cause severe transfusion reaction with hypotension and fever (2). Accordingly, subjects known to contain HLA antibodies should not be used as blood donors.

9.2.2. Platelet Transfusions

Many thrombocytopenic patients can be kept free of bleeding by the infusion of platelet-rich plasma or platelet concentrate. This is particularly true of drug-induced nonimmunological thrombocytopenia, e.g. in patients with malignant diseases treated with cytostatic drugs. In autoimmune thrombocytopenia (for example, in idiopathic thrombocytopenic purpura or in some cases of systemic lupus erythematosus), the autoantibodies are likely to destroy the donor's as well as the patient's own platelets. In the absence of autoantibodies, substitution therapy with random, ABO-compatible platelets is possible. However, when this has been done for some time, most patients eventually develop HLA antibodies, and as these are often multispecific, platelets from random donors are no longer effective because they are destroyed by the antibodies (343). In this situation, it becomes necessary to HLA type the patient and to select platelet donors compatible at least for the antigens of the A and B series. To do this, a large number of blood donors must have been HLA typed in advance, and even then, it is quite often only possible to find a few who are HLA compatible with a given patient. Since at least 3 months must elapse between each blood donation, it is often necessary to perform thrombopheresis: the platelet-rich plasma is given to the patient while the red cells are reinfused in the donor. This procedure can be repeated several times weekly without harming the donor, who must, however, stay in the blood bank for at least an hour each time.

Occasionally, only platelets from HLA-identical siblings survive adequately, and in some cases such donors have maintained the life of their affected relatives for a year or more. However, it should be borne in mind that if an HLA-identical sibling is used as donor, the recipient may become immunized against minor histocompatibility antigens, which might prevent the take of a bone marrow from that donor if this becomes a therapeutic possibility.

9.2.3. Granulocyte Transfusions

It has been shown that the infusion of large numbers of granulocytes may reduce the duration of febrile infection episodes in neutropenic patients (57, 176). Due to the short life span of neutrophils even in normals, the number of granulocytes needed per infusion by far exceeds that of an ordinary unit of blood, and accordingly special measures must be taken to obtain such large yields (57).

Recipients of granulocytes from random donors are, of course, likely to develop HLA antibodies upon continuous therapy, and it has been shown (57) that it is important to match the donors for HLA-AB antigens in such patients in order to get an adequate recovery of the cells infused. Moreover, in such cases the resulting febrile reactions after HLA-incompatible granulocytes are likely to be severe and potentially dangerous due to the high number of cells given. Finally, it seems worth noting that the granulocyte transfusions invariably contain lymphocytes which may perhaps cause graft-versus-host reaction (cf. section 9.1.4.) in recipients receiving immunosuppressive therapy. In such cases, blood products should be irradiated before the transfusion. Irradiation does not interfere with the action of granulocytes.

9.3. HLA in Pregnancy

Pregnancy is nature's own great and successful though time-limited transplantation experiment, and it is still a puzzle why the haploidentical fetus is tolerated without signs of rejection by the maternal organism for 9 months. HLA antigens are present on fetal cells very early (51) and fetal lymphocytes are able to stimulate in the mixed lymphocyte culture test (51). Moreover, the occurrence of alloantibodies against fetal HLA antigens in the sera of about at least a quarter of primigravidae before term demonstrates clearly the immunogenicity of these antigens. However, it is not the purpose to discuss here the various mechanisms which have been suggested to explain the success of this transplantation. This has been done in other reviews (24, 167) and we find it sufficient to mention four possible and often favored explanations: (1) the placenta may serve as a barrier for massive invasion of alloaggressive maternal cells; (2) there is some evidence that the immune function of pregnant women is impaired; (3) the mother may have a specific immunologic enhancement towards the fetus, and (4) the placenta may be devoid of HLA antigens.

As discussed in section 11.1.1. sera from parous women are our main source of complement-fixing lymphocytotoxic antibodies for HLA typing. These antibodies are active against the paternal HLA antigens on fetal cells and as they are usually of IgG immunoglobulin class which passes the placenta, it might be expected that they would be harmful to the fetus. However, except for very rare cases of neonatal thrombocytopenia (cf. below), many large studies have failed to demonstrate definitely increased frequencies of abortions, still births, or developmental anomalies in pregnancies where the mother had HLA antibodies compared to pregnancies without detectable HLA immunization — see Ahrons (1) and Jensen (127) for review.

In extremely rare cases, maternal HLA antibodies can cause platelet destruction in the infant and thus give rise to a so-called isoimmune neonatal purpura (269, 287). The thrombocytopenia is rarely marked at birth but develops within the first few hours and may persist for some weeks. Some children die from intracerebral hemorrhage but most survive without sequelae. Isoimmune neonatal purpura has been estimated to occur once in about 10,000 births (269), and it is worth noting that HLA antibodies seem only to account for a minor part of the cases: most are due to a platelet-specific antibody directed against the so-called Zw^a (or PI^{A1}) antigen present only on platelets. Whichever antibody is involved, the same treatment can be used: packed platelets obtained by thrombopheresis of the mother.

Very recently, evidence has been presented that women who are homozygous at the HLA-A and B loci are more likely than others to develop severe preeclampsia in pregnancy (233).

Before leaving the pregnancy problem we would like to mention a special related topic: *choriocarcinoma*. In this disorder placenta-derived fetal tissues invade the maternal organism as a malignant tumor. Again, we are dealing with nature's allotransplantation experiment but this time with tumor cells. It might be anticipated that this condition is more likely to arise when the fetus is HLA compatible with the mother. Unfortunately, HLA typing of tumor tissue is difficult, but HLA typing of the patients and their husbands have so far not given conclusive evidence that these share HLA-ABC antigens more often than should be expected (168, 195) though women with ABO blood group incompatible husbands are less likely to develop this disorder (168, 194). It has been suggested that the progression of choriocarcinoma depends on the HLA compatibility of the tumor (195), but we feel that this still needs confirmation. Studies of MLC typing of patients and their husbands have not yet been reported.

9.4. Association between HLA and Disease

Perhaps the most fascinating area of HLA research at the present time concerns the observation that various diseases are more likely to occur in individuals carrying certain HLA antigens than in individuals lacking the antigen(s) in question (66, 71, 184, 197, 253, 254, 295).

Studying the frequencies of blood groups in various diseases was first suggested by *Ford* (97) as a means of providing evidence in favor of *Fisher*'s (95) genetic theory of natural selection which has as one of its premises that hardly any gene will be neutral. This prediction was met by the demonstration of *Aird et al.* (4) that gastric cancer occurs significantly more frequently in subjects of blood group A. Since then, an enormous amount of similar data have been published and some other associations found (202, 331). Typically, however, almost all of these associations are weak, and few if any of them have given new hints as to the etiology or pathogenesis of the disorders in question.

In contrast, most of the associations observed between HLA and disease are strong, and several of them have stimulated thinking and experimentation towards better understanding of the disease processes. In fact, one of the associations is so strong that HLA typing is used diagnostically. Obviously, with the knowledge of the biological function of HLA and related animal systems, the perspectives of studying HLA in disease were much better than those for other blood groups, the function of which are as yet largely unknown. For example, the observation that H-2 in mice control the occurrence of viral leukemia (152, 175) and experimental autoimmune thyroiditis (330), and the existence of Ir genes within these genetic systems (183) made HLA a good candidate for disease association studies.

Table XV. Association between HLA-B8 and idiopathic Addison's disease

	HLA-B8		Total
	present	absent	
Patients	22 (58%)	16	38
Controls	467 (23.7%)	1,500	1,967
Total	489	1,516	2,005

Relative risk =

$$\frac{22 \times 1,500}{16 \times 467} = 4.4.$$

$\chi^2 = 23.6$ with 1 degree of freedom, $p < 10^{-5}$. Fisher's exact $p = 10^{-5}$.

9.4.1. How the Studies Are Done

The involvement of HLA in a disease can, in general, be established either by population, by family studies, or by both. The detailed procedures are described in *Armitage* (13), *Li* (174), *Svejgaard et al.* (292) and *Woolf* (339).

(1) *Population studies* are carried out simply by typing a number of un-related patients with the given disease and comparing the frequencies of the various HLA antigens with those observed in a random sample of healthy unrelated controls of the same ethnic origin as the patients. The results for each antigen are usually entered into a 2 × 2 contingency table as illustrated in table XV. It is important to distinguish between the strength and the statistical significance of an association. The strength is conveniently expressed as a *relative risk* which indicates how many times more frequently the disease occurs in a group of individuals carrying the antigen relative to a group lacking it. The relative risk is simply estimated by the cross-product ratio of the four entries in the 2 × 2 table (table XV).

A relative risk higher than one indicates that the frequency of the antigen is increased in the patients, while risks below one indicate decreased frequencies. It is possible to make combined estimates of relative risks from several sources of data even when the frequency of the antigen differs between control groups as may be the case between different ethnic groups (fig. 10). Testing for statistical significance is performed as discussed on page 29. At this juncture, if suffices to say that so-called type I errors (chance deviations) are to be expected because 20 or more different antigens are usually typed for. Type II errors (the 'rejection' of a true association as being due to chance) may also occur, in particular when only a few patients have been investigated. Finally, it is worth noting that it is

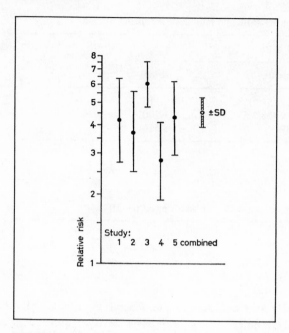

Fig. 10. The ordinate indicates relative risks (log scale) of getting myasthenia gravis for HLA-B8 positives. Estimates of relative risks ± SD are given for five different studies quoted by *Ryder et al.* (254). To the left, the combined estimate ± SD obtained from these five studies is shown. Note that this combination increases the accuracy of the estimate of the risk.

usually more difficult to demonstrate decreased HLA antigen frequencies (relative risks below one) than increased ones (risks above one).

(2) *Family studies.* As will be discussed below, some of the associations observed so far with the HLA-A, B, C, D, and DR antigens may be secondary to primary associations with as yet unknown HLA factors in linkage disequilibrium with the A, B, C and/or D/DR markers in question. If, on the other hand, a disease is associated with an HLA factor not associated with any known markers, none of these would have disturbed frequencies in the patients, and the association would go entirely undetected by population studies of HLA-A, B, C, and D. However, such associations could be revealed by the study of families with more than one member suffering from the disease. Nevertheless, we wish to stress that such studies are likely to meet with considerable difficulties because most diseases of interest in this connection do not have a clear-cut dominant or recessive inheritance: in most cases, environmental differences — and genes at non-HLA linked loci — give rise, e.g. to varying age at onset and even impene-

trance (i.e. some individuals never develop the disease in spite of having the relevant genetic factors).

Family studies are carried out as a kind of linkage studies. Due to the varying age at onset and impenetrance, only affected family members can be considered fully informative, although the rest of the family must often also be typed to allow HLA genotyping. Families with just one affected parent and one affected child provide no information because they inevitably share one HLA haplotype. Accordingly, other affected family members must be available. Families with two affected siblings do provide some, but not much information because three quarters of all sibpairs have one or two HLA haplotypes in common. Thus, at least 20 affected sibpairs must be studied before any reasonable level of significance can be reached if all the pairs share parental haplotypes, if the susceptibility is a dominant trait. If it is recessive, fewer sibpairs are required. *Day and Simons* (73) have developed a special formula for treating data on HLA in affected sibpairs. It should be stressed, however, that families with two affected members probably represent a biased sample as they are likely to possess more disease susceptibility genes than families with only one patient.

Naturally, family studies are also useful in diseases which show association with known HLA markers because they can elucidate the genetics of the disorders and they may also provide evidence that HLA factors different from these markers are involved (cf. below). Moreover, they can give estimates of risks for relatives with and without the disease associated with the HLA factor.

9.4.2. The Associations Found

These are listed in table XVI which gives the relative risks for the antigens showing the most pronounced deviations for the disease in question (253). It is worth noting that almost all of the deviations primarily concern antigens from the B or the D series. Obviously, an increased frequency of, say HLA-B8, from the B series, will cause a 'passive' increase of HLA-A1 from the A series because these two antigens as discussed earlier are positively associated in the population.

In table XVI we have listed mainly the antigens which give the highest risks of the disease in question. Risks below one have been observed but are usually secondary to the increased ones. In most cases, only one antigen is increased, but in psoriasis at least four and in diabetes two antigens show increased risks; as discussed below this situation offers special advantages in the genetic analysis. It seems fair to state that in most cases the increases do not concern entire haplotypes (say *HLA-A1,B17* in psoriasis) but all haplotypes carrying the B series determinant in question (295). The increased risks mean increased susceptibility except, perhaps, in malignancies where they could indicate increased survival value (292).

Table XVI. Association between HLA and disease

Disease	Antigen	Frequency (%) of antigen		Relative risk	Significance p	Number of studies	Number of patients investigated
		controls	patients				
Arthropathies							
Ankylosing spondylitis	B27	9.4	90	87.4	$<10^{-10}$	29	2,022
Reiter's syndrome	B27	9.4	79	37.0	$<10^{-10}$	9	341
Reactive arthritis	B27	increased in yersinia, salmonella, and shigella arthritis					
Psoriatic arthritis	B27	9.4	29	4.0	$<10^{-10}$	5	177
	B16	5.9	15	2.8	$<10^{-3}$	3	138
Juvenile arthritis	B27	9.4	32	4.5	$<10^{-10}$	10	596
Rheumatoid arthritis	Dw4	19.4	50	4.2	$<10^{-9}$	3	152
	DRw4	28.4	70	5.8	$<10^{-5}$	1	53
Eye diseases							
Acute anterior uveitis	B27	9.4	52	10.4	$<10^{-10}$	5	341
Optic neuritis	Dw2	25.8	46	2.4	$<10^{-3}$	2	103
Skin diseases							
Psoriasis vulgaris	B13	4.4	18	4.8	$<10^{-10}$	11	836
	B17	8.0	29	4.8	$<10^{-10}$	11	836
	B37	2.6	11	4.4	$<10^{-6}$	5	323
	Cw6	33.1	87	13.3	$<10^{-10}$	1	40
Dermatitis herpetiformis	Dw3	26.3	85	15.4	$<10^{-10}$	2	61
	DRw3	25	97	56.4	$<10^{-10}$	1	29
Behçet's disease	B5	10.1	41	6.3	$<10^{-10}$	7	147
Intestinal diseases							
Celiac disease	Dw3	26.3	79	10.8	$<10^{-8}$	2	47
Liver diseases							
Chronic autoimmune hepatitis	B8	24.6	75	9.0	$<10^{-10}$	5	161
	DRw3	19	78	13.9	$<10^{-6}$	1	18
'Systemic' diseases							
Bürger's disease	B12	24.4	3	0.1	$<10^{-4}$	2	64
Myasthenia gravis	B8	24.6	57	4.1	$<10^{-10}$	6	328
Sicca syndrome	Dw3	26.3	78	9.7	$<10^{-9}$	2	54
Systemic lupus erythematosus	B8	24.6	41	2.1	$<10^{-10}$	11	554
Hemachromatosis	A3	28.2	76	8.2	$<10^{-10}$	7	191
	B14	3.8	16	4.7	$<10^{-10}$	6	156
	behaves as a recessive trait closely linked to HLA in family studies						

Table XVI (continued)

Disease	Antigen	Frequency (%) of antigen		Relative risk	Significance p	Number of studies	Number of patients investigated
		controls	patients				
Endocrine diseases							
Juvenile and/or insulin dependent diabetes	Dw2	25.8	0	0.0	$<10^{-6}$	2	115
	Dw3	26.3	44	2.2	$<10^{-5}$	2	201
	Dw4	19.4	49	4.0	$<10^{-10}$	2	232
Graves' disease	Dw3	26.3	57	3.7	$<10^{-9}$	3	126
Idiopathic Addison's disease	Dw3	26.3	69	6.3	$<10^{-5}$	1	30
Congenital adrenal hyperplasia	B5	10.1	29	3.6	$<10^{-4}$	2	55
	close linkage to HLA-B in family studies						
Neurologic diseases							
Multiple sclerosis	Dw2	25.8	59	4.1	$<10^{-10}$	9	932
Cerebellar ataxia	perhaps linked to HLA in family studies						
Manic depressive disorder	B16	5.9	13	2.3	$<10^{-4}$	5	313
Allergy							
Ragweed hay fever	B7	25.8	56	3.6	$<10^{-3}$	2	65
	perhaps linked to HLA in family studies						
Infections							
Leprosy	type of leprosy (lepromatous, tuberculoid) apparently linked to HLA in family studies						
Tuberculosis	B8	20	57	5.1	$<10^{-6}$	1	46
Recurrent herpes labialis	A1	32.0	56	2.7	$<10^{-10}$	2	292
Subacute thyroiditis	B35	14.6	70	13.7	$<10^{-10}$	4	105
Malignant diseases							
Hodgkin's disease	A1	32.0	40	1.4	$<10^{-10}$	25	2,669
	B8	24.6	28	1.2	$<10^{-4}$	25	2,670
Acute lymphatic leukemia	A2	53.3	62	1.4	$<10^{-6}$	15	1,099

The above data are from the HLA and Disease Registry (253). The frequencies in controls are mainly from Danish population sample, but when only one study had been reported, the actual figures are given.

9.4.3. Etiology and Pathogenesis

The mere demonstration that a disease shows association with HLA can give important hints as to the etiology and/or pathogenesis of that disorder because we know something about the biological function of HLA. For example, the association between HLA and multiple sclerosis led to immunological investigations which showed that there is a specifically decreased cellular immunity to measles and other paramyxovirus in this disorder (229). Thus, it is possible that multiple sclerosis is the result of a slow virus infection which primarily affects individuals predisposed by an HLA-linked abnormal immune response. In fact, this suggestion has led directly to a clinical trial with transfer factor (an immunostimulatory factor extractable from lymphocytes) in multiple sclerosis (230), but unfortunately this treatment has no influence on the clinical course (96).

The associations between HLA and juvenile diabetes, Graves' disease, and idiopathic Addison's disease lend support to the assumption that these disorders in many cases may be due to autoimmune phenomena – possibly virus-induced – and this has already inspired a number of clinical and experimental studies aiming at clarifying this possibility. The fact that HLA -D3 is associated with a variety of different disorders may indicate a common pathogenetic pathway for at least some of these. The increase of both HLA-D3 and D4 in juvenile diabetes points, as discussed below, to two different predisposing factors each operating by a separate mechanism. The associations between HLA-B27 and ankylosing spondylitis, reactive arthropathies, and acute anterior uveitis tie these three disorders together in an entity clearly separate from rheumatoid arthritis and other arthropathies investigated.

The associations may thus help to distinguish between different variants of a disease. For example, it is clear that the difference between juvenile and maturity onset diabetes in respect of HLA association strongly supports the concept that these two disorders are at least partly different entities. By analogy, the association between HLA and psoriasis vulgaris but not with pustular psoriasis indicates that these are different diseases.

In general, it may be stated that the finding of an association between HLA and a disease points to an infectious and/or autoimmune etiology. However, as pointed out below, other explanations are possible and should be kept in mind.

9.4.4. Diagnostic and Prognostic Implications

So far, only one group of disease has proved so strongly associated with an HLA factor that HLA typing can be claimed to have *diagnostic* value: about 90% of patients with ankylosing spondylitis are B27-positive as compared to only 9%

of healthy individuals. Thus, typing for B27 may be considered a diagnostic test with about 10% false-negative and about 9% false-positive reactions (141). The diagnostic value of this investigation may be illustrated as follows: if there is an a priori chance based on clinical and other data of about 50% that a patient has ankylosing spondylitis, the demonstration that this patient is B27-positive will increase the chance to about 90%, whereas the absence of this antigen decreases the chance to about 10%. However, it should be noted that the B27 antigen is also highly increased in the so-called reactive arthropathies (Reiter's disease and arthritis following yersinia, salmonella, and shigella infections) which may, of course, cause diagnostic problems. On the other hand, the presence of B27 in a patient suffering from juvenile rheumatoid arthritis may indicate a poor prognosis as there is a considerable risk that such patients eventually develop spondylitis and/or acute anterior uveitis, and this seems to be true also of B27-positives suffering from inflammatory bowel disease.

Apart from the reactive arthropathies, where certain prophylactic or thera-peutic measures might be undertaken in B27-positive patients, it seems as if Dw2-positive patients with multiple sclerosis are more likely to have a more rapid progression of their disease than are Dw2-negative patients (130, 229). The clinical significance of this observation is difficult to evaluate at the present time, but when adequate therapies of this disease become available, it is possible that Dw2-positive patients should be treated more intensively than other patients. Similarly, HLA-Dw3-positive patients with Graves' disease are more likely to relapse after treatment with antithyroid drugs (23).

When more becomes known of the underlying mechanisms, it is possible that new prophylactic measures will become available which should of course be offered first of all to the healthy relatives carrying the HLA factor associated with a given disease.

9.4.5. Genetics of Diseases

Most diseases are of a multifactorial nature (48, 49), i.e. they are due to one or often more environmental in addition to genetic factors. Naturally, in some cases, e.g. many infections, the genetic element may be small, but in most cases, the genetic constitution of an individual influences the course of the disease, and some diseases do not occur at all unless the individual possesses the relevant genes. Distinction is usually made between monogenic and polygenic disorders. A *monogenic* disease is due to the action of one gene at one locus and may be either dominant or recessive and may be influenced to varying degrees by environmental factors. Especially, when the environment plays a major role, it can be impossible to decide whether the genetic element is of a monogenic or polygenic nature. *Polygenic* disorders (48) are due to the combined action of genes at many usually not closely linked loci. Each of these genes are thought to bring the individual closer to or past the threshold beyond which the individual is at risk of developing the disease in the appropriate environment. The term *oligogenic* has been suggested to describe disorders which are mainly due to genes at a limited number of loci not closely linked (295). To distinguish between the three kinds of genetic disorders, it is usually necessary to have genetic markers available which are associated with the disease (88) and this is precisely what the HLA system has offered for a variety of diseases (table XVI).

The type of information which can be obtained from studies of HLA in disease may be illustrated with the following examples:

(1) *Ankylosing spondylitis* is a sex-influenced disorder five times as frequent in males as in females (48), and as previously mentioned, it is very strongly, but not absolutely associated with B27 in both sexes and in the four different races (Causasians, Blacks, Japanese, and American Indians) which have been investigated so far. This latter observation would indicate that it is B27 itself which is involved in the pathogenesis and not some other determinant closely linked to B27, because different races seem to show different gametic associations between different HLA factors (133). The fact that about 10% of Caucasian patients lack B27 indicates that one or more genes different from B27 are also involved. It is yet too early to decide whether these additional genes belong to the HLA system or not. However, some interesting pedigrees indicate that they are not HLA linked (77, 243).

There is no evidence that B27-homozygous individuals have a higher risk of developing ankylosing spondylitis than B27-heterozygous ones, which indicates that the susceptibility is purely dominant. *Kidd et al.* (143) and *Thomson and Bodmer* (313) reached the same conclusion using formal genetic analyses of data on HLA-B27 in ankylosing spondylitis.

It seems clear that the *B27* gene or some *B27*-associated gene accounts for a considerable part of the heritability of ankylosing spondylitis which may, therefore, be an example of an oligogenic disorder.

(2) *Psoriasis.* Clinically, distinction has been made between two forms of psoriasis: *psoriasis vulgaris* and *pustular psoriasis.* This distinction now seems to be fundamental as evidenced by the observation that only the former is associated with HLA, while the latter is not. Psoriasis has long been considered a polygenic and multifactorial disorder.

The fact that not less than three HLA antigens (B13, B17, B37) are associated with psoriasis offers special opportunities to study the nature of the association. It appears that there is no increased risk of psoriasis neither for individuals carrying only B13, B17, or B37 from the B series (i.e. probably homozygotes) nor for subjects carrying two of these as compared to those having other B series antigens in addition to one of these three antigens (295).

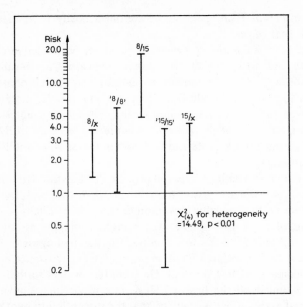

Fig. 11. The bars indicate 95% confidence intervals of relative risks. For simplicity, 'HLA-B' has been omitted from the genotypes. 'X' indicates genotypes including other B antigens than B8 and B15, i.e. truly heterozygous individuals. The homozygosities of HLA-B8/B8 and B15/B15 have not been proven by family studies, but these 'genotypes' indicate absence of B antigens other than B8 and B15, respectively, and these individuals have a high probability of being homozygous. Note that these homozygous genotypes do not confer increased risks as compared to the heterozygotes, i.e. the risk is dominant. In contrast, the HLA-B8/B15 heterozygotes have about twice as high a risk as individuals having only one of the corresponding antigens, i.e. we are faced with an 'overdominance' effect of HLA-B8 and B15. From *Svejgaard et al.* (295), data on juvenile diabetes mellitus.

This indicates that there is no dosage effect for the individual genes and that these in turn have no additive effect on the liability to develop psoriasis. We suggest that this observation can best be interpreted to mean that the three antigens operate through the same mechanism to create increased susceptibility. This assumption is supported by our very recent observation that the increases of B13, B17, and B37 may, in fact, be secondary to a primary increase of the newly defined C series antigen, Cw6, which includes among others each of the above B antigens (211, 255).

A number of families with several members suffering from psoriasis has been HLA-typed, and several of these studies lend support to the assumption that psoriasis can develop without HLA being involved, i.e. that psoriasis is a polygenic disorder.

Interestingly, the effect of HLA seems to be confined mostly to psoriasis starting before the 35th year of age.

(3) The inheritance of *juvenile diabetes mellitus* has never been adequately clarified. Now HLA-B8 and B15 provide two genetic indicators for insulin-dependent diabetes (209, 272, 312). In contrast to the observation in psoriasis, these two factors seem to have a clear additive effect (fig. 11): *B8/B15* hetero-zygotes have about twice the risk of developing insulin-dependent diabetes as have individuals carrying only one of the antigens, whether these be *B8* or *B15* homozygous or not. This indicates that B8 and B15 confer disease susceptibility by two different mechanisms (295, 312).

It now appears that the associations between juvenile diabetes and HLA-B8 and B15 are secondary to 'primary' associations with HLA-Dw3 and Dw4, respectively (208). Moreover, and perhaps even more striking, there seems to be an entire absence of HLA-Dw2 in this disease. Accordingly, it may be assumed that there are three different HLA factors involved in the pathogenesis of juvenile diabetes: an HLA-Dw2-associated factor conferring strong resistance and two other factors associated with Dw3 and Dw4 conferring susceptibility. However, it has been assumed on the basis of HLA data in affected sibpairs, which show a pronounced excess of HLA identity, that the susceptibility may be inherited as a recessive trait (250, 313). If this is true, the HLA gene causing susceptibility must be in strong negative linkage disequilibrium with *Dw2* and in positive disequilibrium with both *Dw3* and *Dw4*. This possibility cannot be ruled out at present, but because the pronounced increased risk of *B8/B15* hetero-zygotes compared to the corresponding homozygotes also seems to exist for the *Dw3/Dw4* heterozygotes as compared to the homozygotes (290), the theory of overdominance is still as likely as that of recessiveness.

9.4.6. Attempts to Explain the Associations

It is outside the scope of this chapter to go into details about these mechanisms but we would like to mention those which have been suggested most often (37, 66, 184, 197, 275, 295, 325).

(1) The fact that most of the associations seem primarily to involve B series or D series antigens has been taken as evidence that the associations are due to the action of specific *immune response (Ir)* genes closely linked to and in linkage disequilibrium with factors from these series. Obviously, Ir genes must be of great biological importance as a defence mechanism against microbial invasions, and the lack of an adequate Ir determinant could give rise to a recessive (as these Ir characters are dominant) susceptibility to certain infections. Conversely, an 'autoaggressive' Ir determinant could give rise to dominant susceptibility to autoimmune phenomena. Moreover, after the discovery of dominant *immune*

suppressor (Is) genes within the H-2 system in mice (29, 301), similar genes in the HLA system may also be responsible for some of the disease associations observed.

(2) In *molecular mimicry* (60, 232) it is assumed that the antigens of certain microorganisms resemble some HLA antigen(s) which would cause a dominant susceptibility to severe infection for individuals carrying these HLA antigens. However, it is also possible that such antigenic resemblance could lead to a breakdown of self-tolerance, and thus to a dominant susceptibility, e.g. to virus-induced autoimmunity.

(3) The above molecular mimicry theory is related to that discussed in section 8.2.2 according to which HLA antigens are *incorporated* in the surface membrane of an infecting virus, which would thus be subject to immunological 'rejection' when infecting a new histoincompatible host.

(4) Cell surface structures such as HLA factors could serve as *receptors* for certain virus which would also cause dominant susceptibility to infection. Some HLA factors could also interfere with the interaction between hormones – or other ligands – and the corresponding receptors on the cell surface; in this way, the HLA system might influence endocrine function or the interaction between cells (296).

(5) Abnormalities of a *complement* factor controlled by the HLA system (such as C2 and Bf) could cause susceptibility to infections or give rise to autoimmune phenomena. In fact, lack of C2 has been found primarily in individuals suffering from lupus-like syndromes (131).

Still other mechanisms are possible but the above seem most likely at the present time.

10. HLA in Paternity Testing and Twin Classification

10.1. Paternity Testing

As soon as HLA typing had reached a level of sufficient reproducibility, it was realized that this very polymorphic system would be of great value in paternity testing. In fact, typing for HLA alone offers about the same amount of information as can be obtained by the investigation of all other genetic markers routinely used in paternity testing concerning the probability of excluding nonbiological fathers (109, 181, 282). Moreover, when the information obtained from all these markers and HLA is pooled, it usually becomes possible to 'prove' with a significant probability that a particular man is indeed the true father.

The use of blood groups and other genetic markers in paternity testing has been excellently discussed by *Race and Sanger* (231), and exactly the same principles apply to the HLA system. In general, this system provides a high rate of exclusion of falsely accused fathers of both of the two classes: a man is excluded (1) if he and the mother lack an antigen which the child has, and (2) if antigens which he must hand on are not present in the child (e.g. an *HLA-A1/A2*-heterozygous father must give either A1 or A2 to his offspring).

Positive evidence in favor of paternity is obtained when a man and the child share an HLA haplotype not found in the mother. It is the fact that most HLA haplotypes are rare, which makes the chance small that another unrelated man could be the true father.

Special reservations must be made when cross-reacting antigens are involved, and the possible synergistic action (cf. p. 84) of two or more antibodies in a typing serum should be kept in mind.

The calculations of probabilities for or against paternity have been given by *Mayr* (181) and *Gürtler et al.* (109) among others.

As HLA typing is a quite expensive method which needs considerable experience, many workers involved in forensic medicine feel that HLA typing should only be used in cases where other less difficult genetic markers do not yield adequate information.

10.2. Twin Zygosity Diagnosis

Twin studies are useful when attempts are made to establish how much inheritance contributes to a disease or a condition because monozygotic twins are genetically identical and differ only by environmental factors, whereas dizygotic twins always show gene differences.

Like other blood groups, the HLA system may help in the distinction between monozygotic and dizygotic twins. Here again, the great polymorphism of HLA ensures that this system is usually informative in such studies. However, the following points should be borne in mind: (1) as 25% of all siblings — including dizygotic twins — are HLA identical, monozygosity can never be proven by HLA typing alone; the maximum weight obtainable by HLA typing in favor of monozygosity is 1/0.25 or 4:1, and accordingly, other genetic systems must be investigated if monozygosity is to be 'proven', and (2) whenever possible, the parents should always be investigated for all the genetic systems, because this provides additional information and greatly facilitates the calculations. These calculations have been developed by *Penrose* and are given in detail by *Race and Sanger* (231).

MLC testing would rarely offer information in addition to that obtained by serological HLA typing.

Chromosome banding would appear to provide much more information in relation to zygosity because the various bands are not closely linked.

As a curiosity it may be mentioned that HLA and blood group chimerism has been observed in dizygotic twins (108, 231). Such rare chimerism arises from the fetal 'transplantation' of bone marrow stem cells through anastomoses in the placenta. Moreover, MLC experiments have shown that such chimeric twins are nonreactive to the 'transplanted' (noninherited) HLA-D antigens (308) in agreement with *Burnet*'s (46) theory of clonal selection.

11. Appendix: Methodology

It is not the purpose of this section to give detailed descriptions of the methods used within HLA work, but we are mainly discussing the principles of these methods and explaining some of the special terms which are often used in the literature dealing with HLA. Detailed descriptions of serological methods for bench workers have been given by *Kissmeyer-Nielsen and Thorsby* (149) and by *Dick and Crichton* (76) and the technical problems of the MLC test have recently been the subject of a workshop (235). References to genetical and statistical procedures are given in sections 3.7 and 9.4.1.

11.1. Classical HLA Typing

11.1.1. Source of Typing Sera

Humoral alloantibodies are the major tools in serological HLA typing. They are found in the sera of individuals immunized through pregnancy, blood transfusions, or by planned immunization with skin grafts and/or injections with leukocytes. A few valuable sera have been obtained by immunizing chimpanzees with human or allologous material, and attempts are made to produce specific HLA antibodies in more remote species such as rabbits by immunization with 'purified' HLA antigens. However, these animals almost invariably form more nonspecific heteroantibodies which must be removed by extensive absorptions before the sera can be used for typing. A promising lead for the future is the possibility to produce large quantities of monoclonal HLA antibodies *in vitro* by fusing antibody-producing cells with mouse myeloma cells. By this method, monoclonal antibodies reacting with an antigen common to all human DR antigens have now been produced (284; *Ferrone,* personal commun.), and the production of a monoclonal anti-HLA-A2 antibody (albeit not a complement-fixing one) has even been reported (219). It should be kept in mind that monoclonality is not synonymous with monospecificity because monoclonal antibodies may well be cross-reacting.

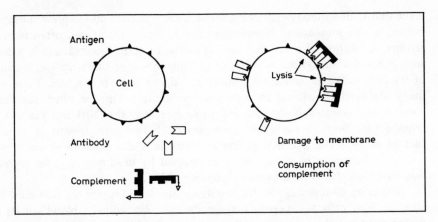

Fig. 12. Schematic presentation of complement fixation induced by the specific binding of complement-fixing antibody to cell surface antigens. The results of this reaction is damage of the cell membrane and consumption of complement. The membrane damage is used in the lymphocytotoxic test, where it is visualized by the uptake of supravital stains such as trypan blue. The consumption of complement is used in the complement-fixation test with platelets.

11.1.2. Typing Procedures

The principle in serological typing is to visualize that an antigen-antibody reaction has taken place. The antibody is contained in the selected immune sera, and the antigen source has so far been the antigens present on the surfaces of cells such as leukocytes, lymphocytes or blood platelets. When mixing a suspension of leukocytes possessing an antigen, say HLA-A1, with a serum containing the corresponding antibody (anti-A1), the cells will often form agglutinates which can be recognized under the microscope. *Leukoagglutination* (144, 241) was the main method in the childhood of HLA typing, but has now been largely abandoned due to technical difficulties often causing poor reproducibility. Instead, another property of the antibodies is now being utilized: their ability to fix complement. The result of this fixation is consumption of complement and damage of the cell membrane carrying the antigen which reacted with the antibody (fig. 12). In the *lymphocytotoxic test* (91, 303), lymphocytes are used as antigen 'carriers', and this test has been almost universally accepted for serological HLA typing at the present time. An almost pure lymphocyte suspension is prepared by density gradient centrifugation of a sample of the peripheral blood, between 1 and 5 ml of blood usually suffice. Routinely, a microtest in special plastic trays is used, i.e. microliter amounts of lymphocyte suspension are incubated with antibody-containing serum under paraffin oil (to prevent evaporation) in the presence of complement (fresh frozen rabbit and often human serum). A solution of a dye such as trypanblue or eosin is added, and it is then seen under the microscope whether or not the cells have taken up dye; staining of the cells indicates membrane damage and thus a positive reaction. There are many technical variants of the lymphocytotoxic test. The one which has been most widely adopted is the two-step procedure (e.g. the 'NIH test') in which lymphocytes and antibody are preincubated before complement is added. Instead of supravital stains (e.g. trypan blue), it is also possible to use labels which are taken up by living cells and released by dead ones, e.g. fluorescein diacetate (39a, 321) or radioactive chromium (238, 259).

The main disadvantage of this lymphocytotoxic test is that the cells must be alive as dead cells are stained 'non-specifically'. Accordingly, blood for HLA typing must usually not be more than a few days old before tested although the addition of certain reagents may prolong this period. Nevertheless, it is possible to freeze and store viable lymphocytes in special media at very low temperatures, for example in liquid nitrogen. For practical purposes, it is worth mentioning that lymphocytes tolerate room temperature (20 °C) better than cooling when stored or shipped unfrozen. Finally, lymphocyte suspensions can also be obtained from lymph nodes or spleens, which is often useful in the cadaver kidney transplantation situation.

Next to the lymphocytotoxic test, the *complement-fixation test with platelets* is the method of choice for serological HLA-ABC typing (55, 269, 287) although it is only carried out at a few laboratories. An almost pure suspension of platelets is made by differential centrifugation of about 2–5 ml of peripheral blood. This test is usually performed as a microtest (55). Equal parts of platelet suspension, antiserum, and complement are mixed and incubated under paraffin oil. Sheep erythrocytes sensitized with anti-sheep-erythrocyte antibody are then added and if these complement-sensitive cells remain intact, this indicates that the complement added initially has been consumed by antibody combining with antigen on the platelets, i.e. a positive reaction. The main advantage of this test is that the platelets need not be alive when typed: they can be stored for at least 3 months at 4 °C before loosing appreciable activity. Another advantage of this test is that the reaction can be read by the naked eye, but as stated earlier, it is considerably more difficult to find sera which can be used for HLA typing in the complement-fixation test with platelets, and accordingly the more sensitive lymphocytotoxic test is the best method for HLA typing at the time of writing.

It is worth noting that both of the methods presently used for HLA typing involves the action of complement, and as some antibody classes are not capable of activating complement, these antibodies would go entirely undetected by these methods. Moreover, most known HLA antibodies belong to the IgG class which only fix complement when two antibody molecules are bound to antigen

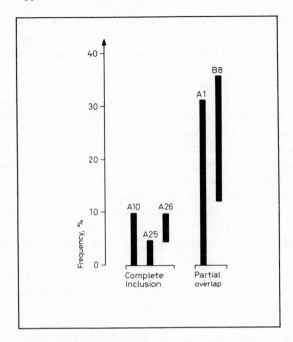

Fig. 13. The inclusion phenomenon. The ordinate indicates the frequency of the population carrying the various antigenic constellations shown at the bottom. Complete inclusions are illustrated to the left: the HLA-A25 and A26 antigens are both completely included in and constitute together the A10 antigen which is defined by an antibody cross-reacting with A25 and A26. The minimal overlap between A25 and A26 is due to *A25/A26* heterozygous individuals. Partial overlap (or inclusion) is illustrated to the right by the occurrence of the A1 and B8 antigens: the B8 antigen is not completely included in A1, but occurs most often together with this antigen because of linkage disequilibrium. In general, complete inclusions are often seen for antigens belonging to the same segregant series, while partial inclusions are characteristic for antigens controlled by genes at neighboring loci, but there are exceptions to this rule.

close to each other. However, as discussed earlier HLA antigens seem to move rather freely in the cell surface of living cells, which may favor the juxtaposition of two antigen-bound antibody molecules.

One of the most sensitive assays for the detection of HLA antibodies is the *antibody-dependent lymphocyte-mediated cytotoxity test* [ADLC; other abbreviations are LALI and ABCIL (78, 326)], which utilizes the fact that the so-called K (killer) cells (which constitute a few percent of the peripheral blood lymphocytes) carry Fc receptors which react with the so-called Fc (constant) part of antigen-bound IgG immunoglobulin and this in turn induces lysis of the 'target' cell carrying the antigen. The test is carried out by ^{51}Cr-labelling the target cells, incubating them with antibody and a surplus of an 'effector' lymphocyte suspension (containing K cells), and then measuring the degree of ^{51}Cr release. This test is 100–1,000 times more sensitive than the ordinary lymphocytotoxic test.

Cross-Reactivity and Other Serological Problems. One of the problems which delayed the unravelling of the genetics of the HLA system derives from the fact that most HLA antibodies are cross-reactive, i.e. the same antibody reacts with more than one antigen because many antigens share antigenic determinants. In addition, many sera contain different antibodies which can be absorbed consecutively with cells carrying the antigens corresponding to the antibody to be removed and which, at the same time, lack the antigen corresponding to the specificity which one wants to keep. In contrast, a cross-reactive antibody directed against, say HLA-A2 and A28, is completely absorbed with cells carrying A2 but lacking A28, and vice versa. Many different constellations of cross-reacting antigens have been observed (56, 212). Some antibodies can even be absorbed with cells which they do not kill in the lymphocytotoxic test or in the agglutination test. Such reactions have been termed *CYNAP* (cytotoxic-negative-absorption-positive) and *ANAP* (agglutination-negative-absorption-positive). For example, some anti-A2 sera are negative in the lymphocytotoxic test with A2-negative, A28-positive cells but the cytotoxic anti-A2 can nevertheless be absorbed with such cells. The most likely explanation for this is that absorption is more sensitive than lymphocytotoxicity.

A *monospecific* antibody reacts only with one antigen, a *duo*(di) specific with two, an *oligo*specific with a few, and a *poly*-(or multi-) specific with several antigenic specificities. It is readily understood that some of the antibodies which we presently consider monospecific may turn out to be duospecific when new sera are found which react with only some of the individuals carrying the 'original' antigen. For example, the HLA-A10 antigen was originally considered to be one antigen, but has now been shown to be *'heterogeneous'* because some sera (containing anti-A25) react only with some A10-positive individuals, whereas others (containing anti-A26) only react with the remaining A10-positives. At the population level, these relationships may give rise to the so-called 'inclusion phenomenon' illustrated in figure 13, which also shows how linkage disequilibrium may cause apparent inclusions.

A serum containing two different antibodies, say anti-A2 and anti-B7, reacts more strongly with cells carrying both of the corresponding antigens (A2 and B7) than with cells carrying only one of them (A2 or B7). This phenomenon is called *synergism* (3, 288), and when the antibodies are weak, it may cause 'false' positive reactions, when such sera are used for HLA typing. By analogy, it is worth noting that monospecific sera have a higher titer with cells homozygous for the gene coding for the corresponding antigen than with heterozygous ones, i.e. HLA genes show *dosage effect* (3, 287).

11.1.3. Typing for HLA-DR Antigens

Naturally, it has been a great challenge to find antibodies directed against HLA-D antigens. Until recently, these efforts, which included immunization procedures, were fruitless but in 1973 *van Leeuwen et al.* (169) described a serum containing an antibody reacting with about 20% of the peripheral blood lymphocytes of some individuals. Moreover, the corresponding antigen seemed to be strongly associated with the HLA-D3 antigen in the population. Unfortunately, the fluorescence method used for the demonstration of this antibody was very delicate and could not be used for routine typing. It is worth noting that at this time it was believed that the MLC antigens were a characteristic of T lymphocytes, but it is now known that these antigens are present primarily on B lymphocytes and monocytes. This point was recently utilized by *van Rood et al.* (246) who used B-cell-enriched lymphocyte suspensions in a lymphocytotoxic test. The B-cell enrichment was obtained by incubating peripheral blood lymphocytes with sheep red blood cells which for unknown reasons stick to T lymphocytes forming so-called 'T rosettes'. These can be removed by flotation and the remaining lymphocytes contain about 70–90% B lymphocytes. In the peripheral blood there are only about 10–20% B lymphocytes, and thus antibodies directed solely against B cells will give only weak and doubtfully positive reactions in the cytotoxic test. With the B-cell-enriched suspensions the distinction between positive and negative reactions is easier. A whole set of B-lymphocyte-specific HLA-DR antigens has now been defined (120), and as mentioned earlier some of these antigens appear to be very strongly associated with the HLA-D antigens, perhaps they are identical to these.

Screening of sera for anti-DR antibodies can also be done with lymphoid cell lines (136) – which are practically always of B-cell origin – or with lymphocytes from patients with chronic lymphatic leukemia (332), which are also B lymphocytes. The simultaneous occurrence of antibodies against the ABC antigens and against B cells obviously creates problems, but it is possible to absorb anti-ABC antibodies by platelets as they lack the DR antigens.

It should be noted that DR typing is still a difficult and laborious procedure and that the results are not always reproducible.

11.1.4. Establishing HLA Tissue Typing

As a kind of synthesis of the above methodological considerations, the principles for establishing an HLA typing laboratory is outlined below. It is worth pointing out that HLA typing is still a delicate procedure which necessitates special training and experience. HLA typing has much in common with blood grouping, and it is mostly useful to develop tissue typing laboratories in close connection with blood grouping laboratories. Antisera against most, but not all HLA antigens may be obtained on request from the NIH serum bank, and some are commercially available. However, most HLA-typing laboratories supply themselves with most, but rarely all, antibodies which they find by *screening* sera from immunized individuals (mostly parous women) against a *panel* of lymphocytes from various donors selected to cover as many different HLA antigens as possible. This panel of donors is an important tool in the laboratory not only for the screening of sera but also for checking the specificity of sera received from other laboratories as minor technical variations may cause the sera to behave differently in various laboratories. Hardly any new typing laboratory can be established within a reasonable time without the collaboration of an experienced laboratory, and even for such laboratories continued mutual exchange of sera and cells is a good guarantee of unchanged quality of the typing. Another satisfying check of the typing is comparison of antigen frequencies in newly typed samples of unrelated individuals with previously typed samples of the same origin. A continuous check of the Hardy-Weinberg equilibrium is valuable in random mating populations.

Special problems arise when special ethnic groups and individuals of other races are to be typed. This is partly due to the fact that some sera which react identically in one population may cease to do so when used in another, and partly because each race appears to have some HLA antigens lacking in others.

11.2. The Mixed Lymphocyte Culture Test

The methodology of the MLC test has been discussed in detail by *Sørensen* (299), in a recent joint report (235), and by *Dupont et al.* (82). The test is based on the fact that lymphocytes carry both *antigens* – HLA in particular – and *receptors* for (foreign) antigens. Thus, when lymphocytes from two non-HLA identical individuals, R and S, are mixed *in vitro,* some of the lymphocytes from R will recognize the HLA antigens of S' lymphocytes as foreign, and conversely, some of S' cells will recognize R's HLA antigens. The fundamental function of a lymphocyte is to divide when it encounters the antigen which it is predestined to recognize (cf. section 9.1.1.), and this is precisely what happens in the MLC test: about 1–2% of the cells undergo a series of divisions, and this process can be estimated by adding ^{14}C- or ^{3}H-labelled thymidine at day 4, 5, or 6 and measuring the degree of isotope-labelling of the cells about 24 h later. Thymidine is used almost exclusively by the cells as a constituent of the chromosomal DNA (deoxyribonucleic acid), i.e. the genetic material as such. When a cell divides, the amount of DNA is duplicated for each division, and thus the amount of thymidine incorporation is a rather accurate relative measure of the number of cells dividing in a given period of time. If the test is carried out as

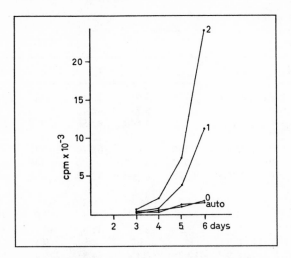

Fig. 14. Kinetics of the one-way MLC test. The abscissa indicates day of 'harvesting' the cultures to which ^{14}C-labelled thymidine was added 24 h earlier. The ordinate indicates cpm as a measure of the uptake of ^{14}C-thymidine. 2 = Stimulation with two-haplotype different cells; 1 = stimulation with haploidentical cells; 0 = stimulation with HLA-identical cells; auto = 'stimulation' with autologous cells. Note (1) the absence of stimulation with HLA-identical cells; (2) the approximate doubling of the stimulation induced by two-haplotype different as compared to haploidentical cells, and (3) the 'exponential' increase of ^{14}C-thymidine incorporation from day 5 to day 6. From *Thomsen et al.* (307).

outlined above, the amount of thymidine incorporation results from divisions of cells from both donors, a so-called *two-way* test. It is possible by pretreatment with certain drugs, usually mitomycin C, or with X-irradiation (about 2,000–5,000 rad) to prevent the cells from dividing, and by mixing untreated cells from R with X-irradiated cells from S (Sx), a *one-way* MLC test is obtained which measures primarily the response of R-*responder* cells to Sx-*stimulating* cells (fig. 3, p. 15).

Figure 14 illustrates the kinetics in the one-way MLC test. It appears that there is an exponential incorporation of thymidine from day 3 to about day 6 of culture, and this in turn indicates exponential cell proliferation. Table XVII gives an example of an MLC experiment. The control is the so-called autostimulated (RRx) or unstimulated (R) 'mixtures'. The degree of stimulation in the allogeneic combination (RSx) is estimated by the counts per minute (cpm) obtained in this mixture minus the cpm in the control. This *increment* of cpm is not always a good estimate of the degree of stimulation, in particular not when responses from various responders are compared. Accordingly, other ways of expressing the results are often used. The *stimulation ratio* or *index* is the cpm in the allogeneic mixture divided by those of the autologous control. This ratio

Table XVII. Various expression of MLC results

Expression	Responder	Stimulator							
		F_x	M_x	1_x	2_x	3_x	4_x	U_x	(D1/D1)$_x$
Counts	father	210	5,120	2,930	2,770	3,010	2,140	5,370	1,400
per minute	mother	9,290	610	3,550	4,730	5,720	4,540	8,280	7,280
	1st child	3,170	4,340	170	250	3,540	8,050	7,920	2,960
	2nd child	2,930	3,250	690	750	2,780	6,520	7,130	2,600
	3rd child	6,740	5,110	7,080	4,370	430	6,370	10,930	9,150
	4th child	3,180	2,250	7,310	6,900	2,790	120	5,740	7,150
	unrelated	6,830	7,170	5,730	6,190	8,300	9,340	200	6,660
Stimulation	father	(1)	24.4	14.0	13.2	14.3	10.2	25.6	6.7
ratio	mother	15.2	(1)	5.8	7.8	9.4	7.4	13.6	11.9
	1st child	18.6	25.5	(1)	1.5	20.8	47.4	46.6	17.4
	2nd child	3.9	4.3	0.9	(1)	3.7	8.7	9.5	3.5
	3rd child	15.7	11.9	16.5	10.2	(1)	14.8	25.4	21.3
	4th child	26.5	18.8	60.9	57.5	23.3	(1)	47.8	59.6
	unrelated	34.2	35.9	28.7	31.0	41.5	46.7	(1)	33.3
Relative	father	(0)	95	53	50	54	37	(100)	23
response, %	mother	113	(0)	38	54	67	51	(100)	87
	1st child	39	54	(0)	1	43	102	(100)	36
	2nd child	34	39	−1	(0)	32	90	(100)	29
	3rd child	60	45	63	38	(0)	57	(100)	83
	4th child	54	38	128	121	48	(0)	(100)	125
	unrelated	98	103	81	88	119	134	(0)	95

The example is identical to that illustrated in a simplified form in table V except for the inclusion of a typing cell (D1/D1)x. *The upper third* gives the 'raw' *cpm*. The cultures are usually set up in triplicate for each combination, and the figures given here could be either means or medians of the triplicate cpm. *The middle third* gives the *stimulation ratios* (SR = SI = stimulation indices) which are calculated for each responder by dividing the cpm (t) of the combination in question with the control cpm (c) for that responder (e.g. the control for the father is FF$_x$ 210 cpm): SR = t/c; e.g. for the combination F1$_x$: SR = 2930/210 = 14.0. *The lower third* gives the *relative responses* (RR) which for each responder is the ratio between the stimulation of the combination in question and the stimulation induced by the unrelated control (U$_x$) which is often a pool of cells. Relative responses are usually given as percentages:

$$RR = \frac{t-c}{U_x-c} \times 100\%; \text{ e.g. for the combination } 4M_x: RR = \frac{2,250-120}{5,740-120} \times 100\% = 38\%.$$

Note the absence of stimulation between siblings 1 and 2, and the 'typing response' of the father and his first two children against the (D1/D1)$_x$ cells. RRs for the unrelated responder are based on the median stimulation [(6,830 + 7,170)/$_2$ = 7,000] excluding autologous and typing cells. More elaborate methods for evaluating MLC response have been given by *Piazza and Galfré* (227) and *Ryder et al.* (256).

seems particularly useful for the definition of *non*-stimulation, whereas it is less meaningful for quantitating definite responses. These are expressed most conveniently as *relative responses,* i.e. the increment counts of the combination in question relative to those of a supposedly 'normal', 'average' or 'median' response of the responder. This normal or '100%' response can, for example, be obtained in the combination RPx where Px is a pool of stimulating cells from about three different unrelated individuals.

As for the lymphocytotoxic test, the lymphocytes for MLC must be viable. This holds not only for the responding but more surprisingly also for the stimulating cells, probably because the antigens of the latter must be presented in a special way which needs protein synthesis. Fortunately, cells stored in liquid nitrogen as discussed above work well both as stimulators and usually also as responders. Refined MLC tests with about 100,000 cells per reaction mixture have now been developed, and it is possible to perform many MLC tests with lymphocytes obtained from 20 or fewer ml of peripheral blood. Several of the steps in the test have been automated, and in general an average laboratory can set up 20 responders in triplicate against 20 stimulators in one day. Apart from being used for HLA-D typing as described below, the MLC test is often applied in cross-matching when living donors are used for transplantation. Even when the donor and recipient are phenotypically HLA-ABC identical siblings, an MLC test should be made to assure that they are genotypically identical and to exclude the possibility that they differ at the D locus due to a recombination between ABC and D or to differences with respect to the paternal gene contribution.

11.2.1. HLA-D Typing (134)

Cells from individuals who are homozygous, say *HLA-D1/D1,* at the HLA-D locus carry only one type, D1, of HLA-D antigen on their surface. When used as stimulators, such cells induce transformation only of responder cells lacking the D1 antigen. Responders possessing D1 themselves cannot recognize the D1 antigen as foreign, and thus will not respond to the *D1/D1* homozygous *'typing cells'.* This is the principle in HLA-D typing as illustrated in table VI (p. 17). The main difficulty is to find homozygous cells. Advantage may, of course be taken of the fact that inbreeding may result in homozygous offspring, and one source of typing cells has been children of first cousin marriages: about $1/_{16}$ of such children are homozygous for one of the four HLA-D determinants of their common great-grand parents. However, several of the MLC antigens have reasonable frequencies in the population and due to the linkage disequilibrium (cf. section 3.7.) between the B locus and the D locus, it is possible to find non-inbred individuals who are HLA-D homozygous by investigating B-homozygous individuals (312).

One of the difficulties in HLA-D typing (once typing cells have been obtained) is caused by the fact that the X-irradiated homozygous, say *D1/D1*, typing cells do recognize the heterozygous, say *D1/D2*, responding cells as foreign and release *blastogenic factor* which nonspecifically stimulates the untreated *D1/D2* cells to divide (back stimulation). Accordingly, a *'typing response'* is rarely a real nonresponse, except when the responders are themselves *D1/D1* homozygous. This is one reason why MLC-typing results are not always unequivocal, and it may, in fact, be difficult to demonstrate that a cell is truly HLA-D homozygous. For example, when the homozygous cells are used as stimulators, responding cells from each of the parents will incorporate more thymidine than the controls due to blastogenic factor released from the stimulating cells. In general, HLA-D typing by means of homozygous cells tend to give false-negative rather than false-positive results, i.e. an HLA-D antigen is occasionally overlooked in the individual who is being typed.

Discovery of methods which could block the action of blastogenic factor without interfering with the MLC reaction would greatly facilitate HLA-D typing. However, the problems of such typing are not really solved until a rapid method is invented.

11.2.2. The Primed Lymphocyte-Typing (PLT) Test

Recently, *Sheehy et al.* (265) described a test which may allow HLA-D typing within 24–48 h. The principle in this elegant test is briefly as follows: an MLC is first made with HLA haploidentical responder and stimulator cells, e.g. from a parent and his offspring with the HLA genotypes, say, *a/b*, and *(a/c)*x. During this primary MLC, the *a/b* cells are stimulated by the HLA-D antigen(s) of the c haplotype. When the culture is allowed to proceed for about 10 days, the proliferation has stopped, but a clone of resting memory (primed) cells reactive against the c antigen is now present. These cells can now be used directly or stored in liquid nitrogen until use. The typing itself is done by mixing the primed cells with the lymphocytes to be typed. If these unknown lymphocytes carry the c antigen, the primed cells will show a rapid secondary response which can be detected by the uptake of isotope-labelled thymidine after 24 h of culture, whereas c-negative cells will only induce the normal MLC response which is first recognizable some days later. A battery of such primed lymphocytes with different specificities can thus be used for PLT typing.

This test is still at an experimental stage, but if it meets with expectations, it represents a significant step forward for the following reasons: (1) it may allow MLC-typing results to be used in the selection of a recipient of a cadaver kidney; (2) it eliminates the necessity of finding the rare HLA-D-homozygous individuals as cell donors for HLA-D typing and (3) as described by the same group of workers (281) recipient lymphocytes primed with a mixture of HLA-D-different cells could be used in a secondary MLC with stimulating donor lymphocytes, which might make it possible to select the most MLC compatible of several possible recipients.

The results obtained so far with the PLT test are somewhat controversial: while some investigators (260) have found that PLT gives results corresponding more closely to HLA-DR than to HLA-D typing, others have observed a good agreement between PLT results and HLA-D typing (311; *Morling and Thomsen,* unpublished). These discrepancies are probably due to differences in the PLT techniques used by the various laboratories.

11.3. Other Methods

Most of the other methods used in HLA research, e.g., biochemical, statistical, and genetical, have been discussed briefly in connection with the sections dealing with the results they have yielded. However, two techniques often applied in the transplantation situation deserve special mention here: the direct and the indirect *cell-mediated lympholysis (CML)* tests.

The *direct* CML test (340) takes advantage of the fact that when an individual is sensitized/immunized through blood transfusion or allotransplantation, a population of killer cells (of T-lymphocyte origin) appears in the blood. These cells are able to react specifically with and lyse cells carrying antigens similar to those of the sensitizing donor. The test is carried out by ^{51}Cr-labelling donor lymphocytes (target cells) and incubating them for about 2–6 h with an excess of lymphocytes from the immunized individual. If there are specific killer (or effector) cells in this blood, these will lyse some of the labelled donor lymphocytes which can be measured by an increase of ^{51}Cr in the supernatant as compared to a control. This test becomes positive shortly after the immunization procedure but remains positive for a few weeks only whereas circulating antibodies may persist for months or years. There are some indications that this test may be used for the detection of a graft rejection episode (107).

In the *indirect* CML (158, 192) the effector cells are generated *in vitro* during an MLC reaction, say ABx. After about 6 days of culture, some A cells have transformed into specific effector cells capable of lysing ^{51}Cr-labelled target lymphocytes from B when mixed with these. Usually the target cells have been kept at special culture conditions from day 0 with the addition of PHA on day 2. Results obtained with this test have provided evidence that the HLA-ABC antigens are the specific targets for the effector cells created in the MLC, whereas the HLA-D antigens necessary to induce the lymphocyte transformation cannot themselves act as targets (90). However, it is possible that some as yet unknown HLA factors may also be targets in the indirect CML test (157).

12. References

1 Ahrons, S.: Human leucocyte antibodies. A brief survey of their properties, occurrence and clinical significance; thesis Århus (1973).

2 Ahrons, S. and Kissmeyer-Nielsen, F.: Febrile transfusion reaction caused by minor specific (LA1) leucocyte incompatibility. Dan. med. Bull. *15:* 257–258 (1968).

3 Ahrons, S. and Thorsby, E.: Cytotoxic HL-A antibodies. Studies of synergism and gene-dose effect. Vox Sang. *18:* 323–333 (1970).

4 Aird, I.; Bentall, H.H., and Roberts, J.A.F.: A relationship between cancer of stomach and the ABO blood groups. Br. med. J. *i:* 799–801 (1953).

5 Albert, E.D.; Mickey, M.R.; Ting, A., and Terasaki, P.I.: Deduction of 2140 HL-A haplotypes and segregation analysis in 535 families. Transplantn Proc. *5:* 215–221 (1973).

6 Albrechtsen, D.; Bratlie, A.; Nousiainen, H.; Solheim, B.G.; Winther, N., and Thorsby, E.: Serological typing of HLA-D: predictive value in mixed lymphocyte cultures (MLC). Immunogenetics *6:* 91–100 (1978).

7 Albrechtsen, D.; Flatmark, A.; Jervell, J.; Halvorsen, S.; Solheim, B.G., and Thorsby, E.: Significance of HLA-D/DR matching in renal transplantation. Lancet *ii:* 1127 (1978).

8 Allen, F.H.: Linkage of HL-A and GBG. Vox Sang. *27:* 382–384 (1974).

9 Amiel, J.F.: Study of the leucocyte phenotypes in Hodgkin's disease. Histocompatibility testing 1967, pp. 79–81 (Munksgaard, Copenhagen).

10 Amos, D.B.; Ruderman, R.; Mendell, N.R., and Johnson, A.H.: Linkage between HL-A and spinal development. Transplantn Proc. *7:* 93–95 (1975).

11 Amos, D.B.: Seigler, H.F.; Southworth, J.G., and Ward, F.E.: Skin graft rejection between subjects genotyped for HL-A. Transplantn Proc. *1:* 342–346 (1969).

12 Appendix. I. The workshop data. Histocompatibility testing 1967, pp. 433–448 (Munksgaard, Copenhagen).

13 Armitage, P.: Statistical methods in medical research (Blackwell, Oxford 1971).

14 Artzt, K. and Bennett, D.: Analogies between embryonic (T/t) antigens and adult major histocompatibility (H-2) antigens. Nature, Lond. *256:* 545–547 (1975).

15 Bach, F.H. and Amos, D.B.: Hu-1: major histocompatibility locus in man. Science, N.Y. *156:* 1506–1508 (1967).

16 Bach, F.H. and Hirschhorn, K.: Lymphocyte interaction. A potential histocompatibility test *in vitro.* Science, N.Y. *143:* 813–814 (1964).

17 Bach, F.H. and Voynow, N.K.: One-way stimulation in mixed leukocyte cultures. Science, N.Y. *153:* 545–547 (1966).

18 Bain, B. and Lowenstein, L.: Genetic studies on the mixed leukocyte reaction. Science, N.Y. *145:* 1315–1316 (1964).

19 Balner, H. and Vreeswijk, W. van: The major histocompatibility complex of Rhesus monkeys (RhL-A). V. Attempts at serological identification of MLR determinants and

postulation of an I region in the RhL-A complex. Transplantn Proc. *7:* suppl. 1, pp. 13–24 (1975).

20 Barnstable, C.J.; Jones, E.A., and Crumpton, M.J.: Isolation, structure and genetics of HLA-A, -B, -C and -DRw (Ia) antigens. Br. med. Bull. *34:* 241–246 (1978).

21 Batchelor, J.R.; Casey, T.A.; Werb, A.; Gibbs, D.C.; Prasad, S.S.; Lloyd, D.F., and James, A.: HLA matching and corneal grafting. Lancet *i:* 551–554 (1976).

22 Batchelor, J.R.; French, M.E.; Cameron, J.S.; Ellis, F., Bewick, M., and Ogg, C.S.: Immunological enhancement of human kidney graft. Lancet *ii:* 1007–1010 (1970).

23 Bech, K.; Lomholtz, B.; Nerup, J.; Thomsen, M.; Platz, P.; Ryder, L.P.; Svejgaard, A.; Siersbæk-Nielsehn, K.; Hansen, J.E.M., and Larsen, J.H.: HLA antigens in Graves' disease. Acta endocr., Copenh. *86:* 510–516 (1977).

24 Beer, A.E. and Billingham, R.E.: Immunobiology of mammalian reproduction. Adv. Immunobiol. *14:* 2–84 (1971).

25 Bekkum, D.W. van: Use and abuse of haemopoietic cell grafts in immune deficiency diseases. Transplantn Rev. *9:* 1–53 (1972).

26 Belvedere, M.C.; Curtoni, E.S.; Dausset, J.; Lamm, L.U.; Mayr, W.; Rood, J. J. van; Svejgaard, A., and Piazza, A.: On the heterogeneity of linkage estimations between LA and FOUR loci of the HL-A system. Tissue Antigens *5:* 99–102 (1975).

27 Belvedere, M.; Mattiuz, P., and Curtoni, E.S.: An antibody cross-reacting with LA and FOUR antigens of the HL-A system. Immunogenetics *1:* 538–548 (1975).

28 Benacerraf, B.: Immune response genes. Scand. J. Immunol. *3:* 381–386 (1974).

29 Benacerraf, B. and Germain, R.N.: The immune response genes of the major histocompatibility complex. Immunol. Rev. *38:* 70–119 (1978).

30 Bergholtz, B.O. and Thorsby, E.: Macrophage-dependent response of immune human T lymphocytes to PPD *in vitro*. Scand. J. Immunol. *6:* 779–786 (1977).

31 Bernoco, D.; Cullen, S.; Scudeller, G.; Trinchieri, G., and Ceppellini, R.: HL-A molecules at the cell surface. Histocompatibility testing 1972, pp. 527–537 (Munksgaard, Copenhagen).

32 Bijnen, A.B.; Schreuder, I.; Volkers, W.S.; Parlevliet, J., and Rood, J.J. van: The lymphocyte activating influence of the HLA-A region. J. Immunogenet. *4:* 1–5 (1977).

33 Binz, H. and Wigzell, H.: Successful induction of specific tolerance to transplantation antigens using autoimmunisation against the recipient's own natural antibodies. Nature, Lond. *262:* 294–295 (1976).

34 Blumenthal, M.; Noreen, H.; Amos, D.B., and Yunis, E.: Genetic mapping of Ir gene in man. Linkage with second locus of HL-A. J. Allergy clin. Immunol. *53:* 93–94 (1974).

35 Bobrow, M.; Bodmer, J.G.; Bodmer, W.F.; McDevitt, H.O.; Lorber, J., and Swift, P.: The search for a human equivalent of the mouse T-locus – negative results from a study of HL-A types in spina bifida. Tissue Antigens *5:* 234–237 (1975).

36 Bodmer, J.G.; Bodmer, W.F., and Piazza, A.: Inclusion analysis of fifth histocompatibility testing workshop sera in 25 populations. Tissue Antigens *5:* 315–366 (1975).

37 Bodmer, W.F.: Evolutionary significance of the HL-A system. Nature, Lond. *237:* 139–145 (1972).

38 Bodmer, W.F.; Cann, H., and Piazza, A.: Differential genetic variability among polymorphisms as an indicator of natural selection. Histocompatibility testing 1972, pp. 753–767 (Munksgaard, Copenhagen).

39a Bodmer, W.; Tripp, M., and Bodmer, J.: Application of a fluorochromatic assay to human leukocyte typing. Histocompatibility testing 1967, pp. 341–350 (Munksgaard, Copenhagen).

39b Boehmer, H. von; Haas, W., and Jerne, N.K.: Major histocompatibility complex-linked immune-responsiveness is acquired by lymphocytes of low-responder mice differentiating in thymus of high-responder mice. Proc. natn. Acad. Sci. USA 75: 2439–2442 (1978).

40 Bourguignon, L.Y.W.; Hyman, R.; Trowbridge, I., and Singer, S.J.: Participation of histocompatibility antigens in capping of molecularly independent cell surface components by their specific antibodies. Proc. natn. Acad. Sci. USA 75: 2406–2410 (1978).

41 Brewerton, D.A.; Caffrey, M.; Hart, F.D.; James, D.C.O.; Nicholls, A., and Sturrock, R.D.: Ankylosing spondylitis and HL-A27. Lancet i: 904–907 (1973).

42 Bubbers, J.E.; Chen, S., and Lilly, F.: Nonrandom inclusion of H-2K and H-2D antigens in Friend virus particles from mice of various strains. J. exp. Med. 147: 340–351 (1978).

43 Buckley, C.E.; Dorsey, F.C.; Corley, R.B.; Ralph, W.B.; Woodbury, M.A., and Amos, D.B.: HL-A-linked human immune-response genes. Proc. natn. Acad. Sci. USA 70: 2157–2161 (1973).

44 Buckner, C.D.; Clift, R.A.; Sanders, J.E.; Williams, B.; Gray, M.; Storb, R., and Thomas, E.D.: ABO-incompatible marrow transplants. Transplantation 26: 233–238 (1978).

45 Burcher, G.W. and Howard, J.C.: A recombinant in the major histocompatibility complex of the rat. Nature, Lond. 266: 362–364 (1977).

46 Burnet, F.M.: Self and non-self (Cambridge University Press, London 1969).

47 Burnet, F.M.: Multiple polymorphism in relation to histocompatibility antigens. Nature, Lond. 245: 359–361 (1973).

48 Carter, C.O.: Genetics of common disorders. Br. med. Bull. 25: 52–57 (1969).

49 Cavalli-Sforza, L.L. and Bodmer, W.F.: The genetics of human populations (Freeman, San Francisco 1971).

50 Ceppellini, R.: A preliminary report on the 3rd International Workshop on Histocompatibility Testing. Adv. Transplantn, pp. 195–202 (Munksgaard, Copenhagen 1968).

51 Ceppellini, R.; Bonnard, G.D.; Coppo, F.; Miggiano, V.C.; Pospisil, M.; Curtoni, E.S., and Pellegrino, M.: Mixed leukocyte cultures and HL-A antigens. I. Reactivity of young fetuses, newborns and mothers at delivery. Transplantn Proc. 3: 58–70 (1971).

52 Ceppellini, R.; Curtoni, E.S.; Mattiuz, P.L.; Miggiano, V.; Scudeller, G., and Serra, A.: Genetics of leukocyte antigens. A family study of segregation and linkage. Histocompatibility testing 1967, pp. 149–185 (Munksgaard, Copenhagen).

53 Ceppellini, R.; Mattiuz, P.L.; Scudeller, G., and Visetti, M.: Experimental allotransplantation in man. I. The role of the HL-A system in different genetic combinations. Transplantn Proc. 1: 385–389 (1969).

54 Cohen, R.J. and Eisen, H.N.: Interactions of macromolecules on cell membranes and restrictions of T-cell specificity by products of the major histocompatibility complex. Cell. Immunol. 32: 1–9 (1977).

55 Colombani, J.; D'amaro, J.; Gabb, J.; Smith, G., and Svejgaard, A.: International agreement on a micro-technique of platelet complement fixation. Transplantn Proc. 3: 121–126 (1971).

56 Colombani, J.; Colombani, M., and Dausset, J.: Crossreactions in the HL-A system with special reference to Da6 cross-reacting group. Description of HL-A antigens Da22, Da23, Da24 defined by platelet complement fixation. Histocompatibility testing 1970, pp. 79–92 (Munksgaard, Copenhagen).

57 Craw, R.G.; Herzig, G.; Perry, S., and Henderson, E.S.: Normal granulocyte transfusion therapy. Treatment of septicemia due to gram-negative bacteria. New Engl. J. Med. 287: 367–371 (1972).

58 Crone, M.; Koch, C., and Simonsen, M.: The elusive T cell receptor. Transplantn Rev. *10:* 36–56 (1972).

59 Curman, B.; Östberg, L.; Sandberg, L.; Malmheden-Eriksson, I.; Stålenheim, G.; Rask, L., and Peterson, P.A.: H-2 linked Ss protein is C4 component of complement. Nature, Lond. *258:* 243–245 (1975).

60 Damian, R.T.: Molecular mimicry: antigen sharing by parasite and host and its consequences. Am. Nat. *98:* 129–134 (1964).

61 Dausset, J.: Leuco-agglutinins. IV. Leuco-agglutinins and blood transfusion. Vox Sang. *4:* 190–198 (1954).

62 Dausset, J.: Iso-leuco-anticorps. Acta haemat. *20:* 156–166 (1958).

63 Dausset, J. and Brecy, H.: Identical nature of leucocyte antigens detectable in monozygotic twins by means of immune iso-leuco-agglutinins. Nature, Lond. *180:* 1430 (1957).

64 Dausset, J.; Colombani, J.; Colombani, M.; Legrand, L. et Feingold, N.: Un nouvel antigène du système HL-A (Hu-1): l'antigène 15 allèle possible des antigènes 1, 11, 12. Nouv. Revue fr. Hémat. *8:* 398–406 (1968).

65 Dausset, J.; Colombani, J.; Legrand, L., and Fellous, M.: Genetics of the HL-A system. Deduction of 480 haplotypes. Histocompatibility testing 1970, pp. 53–75 (Munksgaard, Copenhagen).

66 Dausset, J.; Degos, L., and Hors, J.: The association of the HL-A antigens with diseases. Clin. Immunol. Immunopathol. *3:* 127–149 (1974).

67 Dausset, J. and Hors, J.: HL-A and kidney transplantation. Nature new Biol. *237:* 150–152 (1972).

68 Dausset, J.; Hors, J.; Busson, M.; Festenstein, H.; Oliver, R.T.D.; Pris, A.M.I., and Sachs, J.A.: Serologically defined HL-A antigens and long-term survival of cadaver kidney transplants. A joint analysis of 918 cases performed by France-transplant and the London transplant group. New Engl. J. Med. *290:* 979–984 (1974).

69 Dausset, J.; Ivanyi, P., and Feingold, N.: Tissue alloantigens present in human leukocytes. Ann. N.Y. Acad. Sci. *129:* 486–507 (1966).

70 Dausset, J.; Ivanyi, P., and Ivanyi, D.: Tissue alloantigens in humans. Identification of a complex system (Hu-1). Histocompatibility testing 1965, pp. 51–62 (Munksgaard, Copenhagen).

71 Dausset, J. and Svejgaard, A.: HLA and disease (Munksgaard, Copenhagen 1977).

72 Day, N.K.; Rubinstein, P.; Bracco, M. de; Hansen, J.A.; Good, R.A.; Walker, M.E.; Tulchin, N.; Dupont, B., and Jersild, C.: Hereditary Clr deficiency. Lack of linkage to the HL-A region. Histocompatibility testing 1975, pp. 960–962 (Munksgaard, Copenhagen).

73 Day, N.E. and Simons, M.J.: Disease susceptibility genes – their identification by multiple case studies. Tissue Antigens *8:* 109–119 (1976).

74 Degos, L. et Dausset, J.: Génétique. Association gamétique dans la région chromosomique HL-A (un argument en faveur d'un apport génique par migration). C.r. hebd. Séanc. Acad. Sci., Paris *277:* 2433–2436 (1973).

75 Demant, P.: H-2 gene complex and its role in alloimmune reactions. Transplantn Rev. *15:* 162–200 (1973).

76 Dick, H.M. and Crichton, W.B.: Tissue typing techniques (Williams & Wilkins, Baltimore 1972).

77 Dick, H.M.; Sturrock, R.D.; Goel, G.K.; Henderson, N.; Canesi, B.; Rooney, P.J.; Dick, W.C., and Buchanan, W.W.: The association between HL-A antigens, ankylosing spondylitis and sacro-iliitis. Tissue Antigens *5:* 26–32 (1975).

78 Dickmeiss, E.: Antibody-induced cell-mediated cytotoxicity in an allogeneic human system. Scand. J. Immunol. *2:* 251–260 (1973).

79 Dickmeiss, E.; Søeberg, B., and Svejgaard, A.: Human cell-mediated cytotoxicity against modified target cells is restricted by HLA. Nature, Lond. *270:* 526–528 (1977).

80 Doherty, P.C. and Zinkernagel, R.M.: A biological role for the major histocompatibility antigens. Lancet *i:* 1406–1409 (1975).

81 Dupont, B.; Andersen, V.; Ernst, P.; Faber, V.; Good, R.A.: Hansen, G.S.; Henriksen, K.; Jensen, K.; Juhl, F.; Killmann, S.A.; Koch, C.; Muller-Berat, N.; Park, B.H.; Svejgaard, A.; Thomsen, M., and Wiik, A.: Immunological reconstitution in severe combined immunodeficiency with HL-A incompatible bone marrow graft. Transplantn Proc. *5:* 905–908 (1973).

82 Dupont, B.; Hansen, J.A., and Yunis, E.J.: Human mixed-lymphocyte culture reaction: genetics, specificity, and biological implications. Adv. Immunol. *23:* 107–202 (1976).

83 Dupont, B.; Jersild, C.; Hansen, G.S.; Nielsen, L. Staub; Ryder, L.P.; Thomsen, M., and Svejgaard, A.: The relative importance of SD and LD determinants for the MLC-reaction with special reference to the third SD-series (AJ); in Lymphocyte recognition and effector mechanisms. Proc. 8th Leucocyte Culture Conf., pp. 261–267 (Academic Press, New York 1974).

84 Dupont, B.; Jersild, C.; Hansen, G.S.; Nielsen, L. Staub; Thomsen, M., and Svejgaard, A.: Typing for MLC determinants by means of LD-homozygote and LD-heterozygote test cells. Transplantn Proc. *5:* 1543–1549 (1973).

85 Dupont, B.; Nielsen, L. Staub, and Svejgaard, A.: Relative importance of FOUR and LA loci in determining mixed-lymphocyte reaction. Lancet *ii:* 1336–1340 (1971).

86 Dupont, B.; Oberfield, S.E.; Smithwick, E.M.; Lee, T.D., and Levine, L.S.: Close genetic linkage between HLA and congenital adrenal hyperplasia (21-hydroxylase deficiency). Lancet *ii:* 1309–1312 (1977).

87 Dupont, B.; O'Neill, G.J.; Yang, S.Y.; Pollack, M.S., and Levine, L.S.: Genetic linkage of disease-genes to HLA; in Rose, Bigazzi and Warner, Genetic control of autoimmune disease, pp. 15–25 (Elsevier North-Holland, New York 1978).

88 Edwards, J.H.: Familial predisposition in man. Br. med. Bull. *25:* 58–64 (1969).

89 Ehlers, N. and Ahrons, S.: Cornea transplantation and histocompatibility. Acta ophthal. *49:* 513–527 (1971).

90 Eijsvoogel, V.P.; Bois, M.J.G.J. du; Melief, C.J.M.; Groot-Kooy, M.L. de; Koening, C.; Rood, J.J. van; Leeuwen, A. van; Toit, E. du, and Schellekens, P.T.A.: Position of a locus determining mixed lymphocyte reaction (MLR) distinct from the known HL-A loci, and its relation to cell-mediated lympholysis (CML). Histocompatibility testing 1972, pp. 501–508 (Munksgaard, Copenhagen).

91 Engelfriet, C.P.: Cytotoxic isoantibodies against leucocytes; thesis Aemstelstad (1966).

92 Ettinger, R.B.; Terasaki, P.I.; Opelz, G.; Malekzadeh, M.; Pennisi, A.J.; Uittenbogaart, C., and Fine, R.: Successful renal allografts across a positive cross-match for donor B-lymphocyte alloantigen. Lancet *ii:* 56–58 (1976).

93 Falchuk, Z.M.; Rogentine, G.N., and Strober, W.: Predominance of histocompatibility antigen HL-A8 in patients with gluten-sensitive enteropathy. J. clin. Invest. *51:* 1602–1605 (1972).

94 Festenstein, H.; Sachs, J.A., and Oliver, R.T.D.: Genetic studies of the mixed lymphocyte reaction in H-2 identical mice; in Immunogenetics of the H-2 system, pp. 170–177 (Karger, Basel 1971).

95 Fisher, R.A.: The genetical theory of natural selection (Oxford University Press, London 1930).

96 Fog, T.; Pedersen, L.; Raun, N.E.; Kam-Hansen, S.; Mellerup, E.; Platz, P.; Ryder, L.P.; Jakobsen, B.K., and Grob, P.: Long-term transfer-factor treatment for multiple sclerosis. Lancet *i:* 851–853 (1978).

97 Ford, E.B.: Polymorphism. Biol. Rev. *20:* 73–88 (1945).

98 Francke, U. and Pellegrino, M.A.: Assignment of the major histocompatibility complex to a region of the short arm of human chromosome 6. Proc. natn. Acad. Sci. USA *74:* 1147–1151 (1977).

99 Fu, S.M.; Kunkel, H.G.; Brusman, H.P.; Allen, F.H., and Fotino, M.: Evidence for linkage between HL-A histocompatibility genes and those involved in the synthesis of the second component of complement. J. exp. Med. *140:* 1108 (1974).

100 Fu, S.M.; Stern, R.; Kunkel, H.G.; Dupont, B.; Hansen, J.A.; Day, N.K.; Good, R.A.; Jersild, C., and Fotino, M.: MLC-determinants and C2 deficiency: LD-7a associated with C2 deficiency in four families. J. exp. Med. *142:* 495–506 (1975).

101a Gatti, R.A.; Meuwissen, H.J.; Terasaki, P.I., and Good, R.A.: Recombination within the HL-A locus. Tissue Antigens *1:* 239–241 (1971).

101b Giles, C.M.; Gedde-Dahl, T.; Robson, E.B.; Thorsby, E.; Olaisen, B.; Arnason, A.; Kissmeyer-Nielsen, F., and Schrueder, I.: Rgᵃ (Rodgers) and the HLA region: linkage and associations. Tissue Antigens *8:* 143–149 (1976).

102 Goodfellow, P.N.; Jones, E.A.; Heyningen, V. van; Solomon, E.; Bobrow, M.; Miggiano, V., and Bodmer, W.F.: The β_2-microglobulin is on chromosome 15 and not in the HL-A region. Nature, Lond. *254:* 267–269 (1975).

103 Goulmy, E.; Termijtelen, A.; Bradley, B.A., and Rood, J.J. van: Y-antigen killing by T cells of women is restricted by HLA. Nature, Lond. *266:* 544–545 (1977).

104 Green, I.: Genetic control of immune responses. Immunogenetics *1:* 4–21 (1974).

105 Griepp, R.B.; Stinson, E.B.; Oyer, P.E.; Bieber, C.; Dong, E., and Shumway, N.E.: Management of long-term survivors of heart transplantation. Transplantn Proc. *7:* suppl. 1, pp. 595–599 (1975).

106 Grosse-Wilde, H.; Weil, J.; Alberg, E.; Scholz, S.; Bidlingmaier, F.; Sippell, W.G., and Knorr, D.: Genetic linkage studies between congenital adrenal hyperplasia and the HLA blood group system. New Engl. J. Med. (submitted).

107 Grunnet, N.; Kristensen, T.; Kornerup, H.J., and Kissmeyer-Nielsen, F.: Direct cell mediated lympholysis. A test of an allograft-rejection in human kidney recipients. Tissue Antigens *5:* 280–285 (1975).

108 Gundolf, F. and Hansen, H.E.: Lymphocyte and HL-A chimerism in a pair of blood group chimeric twins. Acta path. microbiol. scand. *80:* 152–154 (1972).

109 Gürtler, H.; Svejgaard, A.; Nielsen, L. Staub, and Thomsen, J.: The HL-A system in cases of disputed paternity. Exclusions and positive evidence for paternity; in Gesellschaft für forensische Blutgruppenkunde, Kongressbericht 1973.

110 Hammerberg, C. and Klein, J.: Linkage disequilibrium between H-2 and t complexes in chromosome 17 of the mouse. Nature, Lond. *258:* 296–299 (1975).

111 Hansen, G.S.; Rubin, B., and Sørensen, S.F.: Human leucocyte responses *in vitro*. I. Transformation of purified T lymphocytes with and without addition of partially purified monocytes. Clin. exp. Immunol. *29:* 295–303 (1977).

112 Hansen, G.S.; Rubin, B.; Sørensen, S.F., and Svejgaard, A.: Importance of HLA-D antigens for the cooperation between human monocytes and T lymphocytes. Eur. J. Immunol. *7:* 520–525 (1978).

113 Hansen, H.E. and Eriksen, B.: HLA-GLO linkage analysis in 57 informative families with a total of 145 children, and a study of linkage equilibrium. Hum. Hered. (submitted).

114 Hansen, H.E.; Ryder, L.P., and Nielsen, L. Staub: Recombination between the second and third series of the HL-A system. Tissue Antigens *6:* 275–277 (1975).

115 Hansen, H.E.; Spärck, J.V., and Larsen, S.O.: An examination of HLA frequencies in three age groups. Tissue Antigens *10:* 49–55 (1977).

116 Harris, R. and Zervas, J.D.: Reticulocyte HL-A antigens. Nature, Lond. *221:* 1062–1063 (1969).

117 Hess, M. and Smith, W.: Comparative studies on mouse (H-2) and human (HLA) histocompatibility antigens. Eur. J. Biochem. *43:* 471–477 (1974).

118 Histocompatibility testing 1972 (Munksgaard, Copenhagen).

119 Histocompatibility testing 1975 (Munksgaard, Copenhagen).

120 Histocompatibility testing 1977 (Munksgaard, Copenhagen).

121 Horowitz, S.D.; Groshong, T.; Bach, F.H.; Hong, R., and Yunis, E.J.: Treatment of severe combined immunodeficiency with bone-marrow from an unrelated, mixed-leucocyte-culture-non-reactive donor. Lancet *ii:* 431–433 (1975).

122a Horowitz, S.D. and Hong, R.: The pathogenesis and treatment of immunodeficiency. Monogr. Allergy, vol. 10 (Karger, Basel 1977).

122b Immunological Reviews, vol. 38: Ir genes and T lymphocytes (1978).

123a Immunological Reviews, vol. 40: Role of macrophages in the immune response (1978).

123b Immunological Reviews, vol. 42: Acquisition of the T cell repertoire (1978).

124 Ivanyi, P.: Some aspects of the H-2 system, the major histocompatibility system in the mouse. Proc. R. Soc. Lond. B *202:* 117–158 (1978).

125 Ivanyi, P. and Forejt, J.: Genetic factors closely associated with the major histocompatibility system. Structural and/or regulatory genes. Proc. 2nd Meet. European and African Division of the International Society of Haematology. Excerpta med. *1974:* 143–152.

126 Janeway, C.A.; Wigzell, H., and Binz, H.: Two different V_H gene products make up the T-cell receptors. Scand. J. Immunol. *5:* 993–1001 (1976).

127 Jensen, K.G.: Leucocyte antibodies and pregnancy. A survey; thesis Copenhagen (1966).

128 Jerne, N.K.: The somatic generation of immune recognition. Eur. J. Immunol. *1:* 1–9 (1971).

129 Jerne, N.K.: Towards a network theory of the immune system. Ann. Immunol., Paris *125c:* 373–389 (1974).

130 Jersild, C.; Fog, T.; Hansen, G.S.; Thomsen, M.; Svejgaard, A., and Dupont, B.: Histocompatibility determinants in multiple sclerosis with special reference to clinical course. Lancet *ii:* 1221–1225 (1973).

131 Jersild, C.; Rubinstein, P., and Day, N.K.: Complement and the major histocompatibility systems; in Day and Good, Biological amplification systems in immunology, pp. 247–275 (Plenum Press, New York 1977).

132 Joint report of the Fourth International Histocompatibility Workshop. Histocompatibility testing 1970, pp. 17–44 (Munksgaard, Copenhagen).

133 Joint report of the Fifth International Histocompatibility Workshop. Histocompatibility testing 1972, pp. 618–778 (Munksgaard, Copenhagen).

134 Joint report of the Sixth International Histocompatibility Workshop Conference. II. Typing for HLA-D (LD-1 or MLC) determinants. Histocompatibility testing 1975, pp. 414–458 (Munksgaard, Copenhagen).

135 Joint report, A,B,C of HLA. Histocompatibility testing 1975, pp. 21–99 (Munksgaard, Copenhagen).

136 Jones, E.A.; Goodfellow, P.N.; Bodmer, J.G., and Bodmer, W.F.: Serological identification of HL-A-linked human 'Ia-type' antigens. Nature, Lond. *256:* 650–652 (1975).

137 Jørgensen, F.; Lamm, L.U., and Kissmeyer-Nielsen, F.: Mixed lymphocyte culture with inbred individuals. An approach to MLC typing. Tissue Antigens *3:* 323–339 (1973).

138 Jørgensen, F.; Lamm, L.U., and Kissmeyer-Nielsen, F.: Three LD (MLC) determinants. A Danish population study. Tissue Antigens *4:* 419–428 (1974).

139 Kasakura, S. and Lowenstein, L.: The effect of irradiation *in vitro* on mixed leukocyte cultures with phytohaemagglutinin. Histocompatibility testing 1965, pp. 211–212 (Munksgaard, Copenhagen).

140 Katz, D.H.; Hamaoka, T., and Benacerraf, B.: Cell interaction between histocompatible T and B lymphocytes. II. Failure of physiologic cooperative interactions between T and B lymphocytes from allogeneic donor strains in humoral responses to hapten-protein conjugates. J. exp. Med. *137:* 1404–1429 (1973).

141 Katz, M.A.: A probability graph describing the predictive value of a highly selective diagnostic test. New Engl. J. Med. *291:* 1115–1116 (1974).

142 Keuning, J.J.; Tweel, J.G. van den; Gabb, B.W.; Termijtelen, A.; Goulmy, E.; Blokland, E.; Elferink, B.G., and Rood, J.J. van: An estimation of the recombination fraction between the MLC locus and FOUR locus. Tissue Antigens *6:* 107–115 (1975).

143 Kidd, K.K.; Bernoco, D.; Carbonara, A.O.; Daneo, V.; Steiger, U., and Ceppellini, R.: Genetic analysis of HLA associated diseases: The 'illness susceptible' gene frequency and sex ratio in ankylosing spondylitis; in Dausset and Svejgaard, HLA and disease, pp. 72–80 (Munksgaard, Copenhagen 1977).

144 Killmann, S.-Å.: Leucocyte agglutinins, properties, occurrence and significance; thesis Oxford (1960).

145 Kissmeyer-Nielsen, F.; Ehlers, N.; Kristensen, T., and Lamm, L.U.: The HLA system, serology and transplantation; in Ferrara, HLA system – new aspects, pp. 69–91 (Elsevier North-Holland, Amsterdam 1977).

146 Kissmeyer-Nielsen, F.; Svejgaard, A., and Hauge, M.: Genetics of the human HL-A transplantation system. Nature, Lond. *219:* 1116–1119 (1968).

147 Kissmeyer-Nielsen, F.; Svejgaard, A., and Nielsen, L. Staub: Crossing over within the HL-A system. Nature, Lond. *224:* 75–76 (1969).

148 Kissmeyer-Nielsen, F.; Sørensen, S.F.; Svejgaard, A.; Nielsen, L. Staub, and Thorsby, E.: Mixed lymphocyte cultures and HL-A identity in unrelated subjects. Nature, Lond. *228:* 63–65 (1970).

149 Kissmeyer-Nielsen, F. and Thorsby, E.: Human transplantation antigens. Transplantn Rev. *4:* 1–176 (1970).

150 Klareskog, L.; Rask, L.; Fohlman, J., and Peterson, P.A.: Heavy HLA-DR (Ia) antigen chain is controlled by the MHC region. Nature, Lond. *275:* 762–764 (1978).

151 Klareskog, L.; Tjernlund, U.M.; Forsum, U., and Peterson, P.A.: Epidermal Langerhans cells express Ia antigens. Nature, Lond. *268:* 249–250 (1977).

152 Klein, J.: Biology of the mouse histocompatibility-2 complex (Springer, New York 1975).

153 Klein, J.; Flaherty, L.; VandeBerg, J.L., and Shreffler, D.C.: H-2 haplotypes, genes, regions, and antigens: first listing. Immunogenetics *6:* 459–512 (1978).

154 Koch, C.T.; Hooff, J.P. van; Leeuwen, A. van; Tweel, J.G. van den; Frederiks, E.; Steen, G.J. van der; Schippers, H.M.A., and Rood, J.J. van: The relative importance of matching for the MLC versus the HL-A loci in organ transplantation. Histocompatibility testing 1972, pp. 521–524 (Munksgaard, Copenhagen).

155 Kourilsky, F.M.; Dausset, J.; Feingold, N.; Duprey, J.M., and Bernard, J.: Leucocyte groups and acute leukemia. J. natn. Cancer Inst. *41:* 81–87 (1968).

156 Kourilsky, F.M.; Silvestre, D.; Neauport-Sautes, C.; Loosfelt, Y., and Dausset, J.: Antibody-induced redistribution of HL-A antigens at the cell surface. Eur. J. Immunol. *2:* 249–257 (1972).

157 Kristensen, T.; Grunnet, N., and Kissmeyer-Nielsen, F.: Indirect cell mediated lympholysis (ICML) in man. Evidence for a separate CML locus? Histocompatibility testing 1975, pp. 835–844 (Munksgaard, Copenhagen).

158 Kristensen, T.; Grunnet, N., and Kissmeyer-Nielsen, F.: Cell mediated lympholysis in man. Varying strength of the HL-A (LA and FOUR) antigens as sensitizing or target determinants. Tissue Antigens 6: 221–228 (1975).

159 Kristensen, T. and Jørgensen, F.: False HLA-D assignments may be caused by cytotoxic responder lymphocytes. Tissue Antigens 11: 443–448 (1978).

160 Lachmann, P.J.; Grennan, D.; Martin, A., and Demant, P.: Identification of Ss protein as murine C4. Nature, Lond. 258: 242–243 (1975).

161 Lachmann, P.J. and Hobart, M.J.: Complement genetics in relation to HLA. Br. med. Bull. 34: 247–252 (1978).

162 Lamm, L.U.; Friedrich, U.; Petersen, G.B.; Jørgensen, J.; Nielsen, J.; Therkelsen, A.J., and Kissmeyer-Nielsen, F.: Assignment of the major histocompatibility complex to chromosome No. 6 in a family with a pericentric inversion. Hum. Hered. 24: 243–284 (1974).

163 Lamm, L.U.; Kissmeyer-Nielsen, F.; Svejgaard, A.; Petersen, G.B.; Thorsby, E.; Mayr, W., and Högman, C.: On the orientation of the HL-A region and the PGM_3 locus in the chromosome. Tissue Antigens 2: 205–214 (1972).

164 Lamm, L.U.; Svejgaard, A., and Kissmeyer-Nielsen, F.: PGM_3:HL-A, a new linkage in man. Nature new Biol. 231: 109–110 (1971).

165 Lamm, L.U.; Thorsen, I.-L.; Petersen, G.B.; Jørgensen, J.; Henningsen, K.; Bech, B., and Kissmeyer-Nielsen, F.: Data of the HL-A linkage group. Ann. hum. Genet. 38: 383–390 (1975).

166 Lamm, L.U.; Weitkamp, L.R.; Jensson, O.; Pedersen, G.B., and Kissmeyer-Nielsen, F.: On the mapping of PGM_3, GLO and HLA. Tissue Antigens 11: 132–138 (1978).

167 Lancet: the fetus as a homograft. Lancet i: 535–536 (1975).

168 Lawler, S.L.; Klouda, P.T., and Bagshawe, K.D.: The HL-A system in trophoblastic neoplasia. Lancet ii: 834–837 (1971).

169 Leeuwen, A. van; Schutt, H.R.E., and Rood, J.J. van: Typing for MLC (LD). II. The selection of non stimulator cells by MLC inhibition tests using SD identical stimulator cells (MISIS) and fluorescent antibody studies. Transplantn Proc. 5: 1539–1542 (1973).

170 Legrand, L. and Dausset, J.: The complexity of the HL-A gene product. II. Possible evidence for a 'public' determinant common to the first and second HL-A series. Transplantation 19: 177–180 (1975).

171 L'Esperance, P.; Hansen, J.A.; Jersild, C.; O'Reilly, R.; Good, R.A.; Thomsen, M.; Nielsen, L. Staub; Svejgaard, A., and Dupont, B.: Bone marrow donor selection among unrelated FOUR-locus identical individuals. Transplantn Proc. 7: 823–831 (1975).

172 Levine, B.B.; Stember, R.H., and Fotino, M.: Ragweed hay fever: genetic control and linkage to HL-A haplotypes. Science, N.Y. 178: 1201–1203 (1972).

173 Lewis, M.; Kaita, H.; Chown, B.; Bowen, P.; Lee, C.S.N.; McDonald, S.; Giblett, E.R.; Anderson, J.; Dossetor, J.B.; Schlaut, J.; Pai, K.R.M.; Singal, D.P., and Steinberg, A.G.: A genetic linkage analysis of chromosome 6 markers Chido, HLA and glyoxalase; in Human Gene Mapping 3. Baltimore Conference (1975). Birth Defects Orig. Article Ser., vol. XII, No. 7, pp. 317–321 (1976).

174 Li, C.C.: Human genetics – principles and methods (McGraw-Hill, New York 1961).

175 Lilly, F.; Boyse, E.A., and Old, L.J.: Genetic basis of susceptibility to viral leukaemogenesis. Lancet ii: 1207–1209 (1964).

176 Lowenthal, R.M.; Goldman, J.M.; Buskard, N.A.; Murphy, B.C.; Grossman, L.; Storring, R.A.; Park, D.S.; Spiers, A.S.D., and Galton, D.A.G.: Granulocyte transfusions in treatment of infections in patients with acute leukaemia and aplastic anaemia. Lancet i: 353–358 (1975).

177 Löw, B.; Messeter, L.; Månsson, S., and Lindholm, T.: Crossing-over between the SD-2 (FOUR) and SD-3 (AJ) loci of the human major histocompatibility chromosomal region. Tissue Antigens 4: 405 (1974).

178 Mann, D.L.: The effect of enzyme inhibitors on the solubilization of HLA antigens with 3 M KCl. Transplantation 14: 398–401 (1972).

179 Martins-da-Silva, B.; Vassalli, P., and Jeannet, M.: Matching renal grafts. Lancet i: 1048 (1978).

180 Mattiuz, P.L.; Ihde, D.; Piazza, A.; Ceppellini, R., and Bodmer, W.F.: New approaches to the population genetic and segregation analysis of the HL-A system. Histocompatibility testing 1970, pp. 193–205 (Munksgaard, Copenhagen).

181 Mayr, W.R.: Grundlagen zur Berechnung der Vaterschaftswahrscheinlichkeit im HL-A System. Z. ImmunForsch. exp. Ther. 144: 18–27 (1972).

182 Mayr, W.R.; Bernoco, D.; Marchi, M. de, and Ceppellini, R.: Genetic analysis and biological properties of products of the third SD(AJ) locus of the HL-A region. Transplantn Proc. 5: 1581–1593 (1973).

183 DcDevitt, H.O. and Benacerraf, B.: Genetic control of specific immune response. Adv. Immunol. 11: 31–74 (1969).

184 McDevitt, H.O. and Bodmer, W.F.: HL-A, immune response genes and disease. Lancet i: 1269–1275 (1974).

185 McDevitt, H.O. and Tyan, M.L.: Genetic control of the antibody response in inbred mice. J. exp. Med. 128: 1–11 (1968).

186 McKusick, V.A.: Mendelian inheritance in man; 4th ed. (The John Hopkins University Press, Baltimore 1975).

187 McMichael, A.J. and Sasazuki, T.: A suppressor T cell in the human mixed lymphocyte reaction. J. exp. Med. 146: 368–380 (1977).

188 McMichael, A.J.; Ting, A.; Zweerink, H.J., and Askonas, B.A.: HLA restriction of cell-mediated lysis of influenza virus-infected human cells. Nature, Lond. 270: 524–526 (1977).

189 Mempel, W.; Grosse-Wilde, H.; Baumann, P.; Netzel, B.; Steinbauer-Rosenthal, I.; Scholz, S.; Bertrams, J., and Albert, E.D.: Population genetics of the MLC response: typing for MLC determinants using homozygous and heterozygous reference cells. Transplantn Proc. 5: 1529–1534 (1973).

190 Meruelo, D. and Edidin, M.: Association of mouse liver adenosine 3′,5′-cyclic monophosphate (cyclic AMP) levels with histocompatibility-2 genotype. Proc. natn. Acad. Sci. USA 72: 2644–2648 (1975).

191 Middleton, J.; Crookston, M.C.; Falk, J.A.; Robson, E.B.; Cook, P.J.K.; Batchelor, J.R.; Bodmer, J.; Ferrara, G.B.; Festenstein, H.; Harris, R.; Kissmeyer-Nielsen, F.; Lawler, S.D.; Sachs, J.A., and Wolf, E.: Linkage of Chido and HL-A. Tissue Antigens 4: 366–373 (1974).

192 Miggiano, V.C.; Bernoco, D.; Lightbody, J.; Trinchieri, G., and Ceppellini, R.: Cell-mediated lympholysis in vitro with normal lymphocytes as target. Specificity and cross-reactivity of the test. Transplantn Proc. 4: 231–237 (1972).

193 Miller, J.F.A.P. and Vadas, M.A.: The major histocompatibility complex: influence on immune reactivity and T-lymphocyte activation. Scand. J. Immunol. 6: 771–778 (1977).

194 Mittal, K.K.; Kachru, R.B.; Kahan, B.D., and Brewer, J.J.: The HL-A and ABO antigens in throphoblastic neoplasia. Tissue Antigens 6: 57–69 (1975).

195 Mogensen, B. and Kissmeyer-Nielsen, F.: Invasive mole, gestational choriocarcinoma and transplantation antigens. Ser. haematol. 5: 22–43 (1972).

196 Mollison, P.L.: Blood transfusion in clinical medicine; 5th ed. (Blackwell, Oxford 1972).

197 Morris, P.J.: Histocompatibility systems, immune response and disease in man. Contemp. Topics Immunobiol., vol. 3, pp. 141–169 (Plenum Press, New York 1974).

198 Morris, P.J.; Lawler, S.D., and Oliver, P.T.: HL-A and Hodgkin's disease. Histocompatibility testing 1972, pp. 669–677 (Munksgaard, Copenhagen).

199 Morris, P.J.; Ting, A.; Oliver, D.O.; Bishop, M.; Williams, K., and Dunnill, M.S.: Renal transplantation and a positive serological cross-match. Lancet *i:* 1288–1291 (1977).

200 Morton, J.A.; Pickles, M.M., and Sutton, L.: The correlation of the Bg^a blood group with the HL-A7 leucocyte group. Demonstration of antigenic sites on red cells and leucocytes. Vox Sang. *17:* 536–547 (1969).

201 Morton, J.A.; Pickless, M.M.; Sutton, L., and Skov, F.: Identification of further antigens on red cells and lymphocytes. Vox Sang. *21:* 141–153 (1971).

202 Mourant, A.E.; Kopec, A.C., and Domaniewska-Sobczak, K.: Blood groups and diseases. A study of associations of diseases with blood groups and other polymorphisms. Oxford monographs on medical genetics (Oxford University Press, Oxford 1978).

203 Munro, A.J. and Taussig, M.J.: Two genes in the major histocompatibility complex control immune response. Nature, Lond. *256:* 103–106 (1975).

204 Müller-Eberhard, H.J.: Chemistry and reaction mechanism of complement. Adv. Immunol. *8:* 1–80 (1968).

205 Möller, G.: T and B lymphocytes in humans. Transplantn Rev. *16* (1972).

206 Nandi, S.: The histocompatibility-2 locus and susceptibility to Bittner virus borne by red blood cells in mice. Proc. natn. Acad. Sci. USA *58:* 485–492 (1967).

207 Nathenson, S.G. and Cullen, S.E.: Biochemical properties and immunochemical genetic relationships of mouse H-2 alloantigens. Biochim. biophys. Acta *344:* 1–25 (1974).

208 Nerup, J.; Cathelineau, C.; Seignalet, J., and Thomsen, M.: HLA and endocrine diseases; in Dausset and Svejgaard, HLA and disease, pp. 149–167 (Munksgaard, Copenhagen 1977).

209 Nerup, J.; Platz, P.; Andersen, O.O.; Christy, M.; Lyngsøe, J.; Poulsen, J.E.; Ryder, L.P.; Nielsen, L. Staub; Thomsen, M., and Svejgaard, A.: HL-A antigens and diabetes mellitus. Lancet *ii:* 864–866 (1974).

210 Nielsen, L. Staub; Jersild, C.; Ryder, L.P., and Svejgaard, A.: HL-A antigen, gene, and haplotype frequencies in Denmark. Tissue Antigens *6:* 70–76 (1975).

211 Nielsen, L. Staub; Ryder, L.P., and Svejgaard, A.: The third (AJ) segregant series. Histocompatibility testing 1975, pp. 324–329 (Munksgaard, Copenhagen).

212 Nielsen, L. Staub and Svejgaard, A.: HL-A immunization and HL-A types in pregnancy. Tissue Antigens *2:* 316–327 (1972).

213 O'Neill, G.J.; Yang, S.Y.; Tegoli, J.; Berger, R., and Dupont, B.: Chido and Rodgers blood groups are distinct antigenic components of human complement C4. Nature, Lond. *273:* 668–670 (1978).

214 Opelz, G. and Terasaki, P.I.: Improvement of kidney-graft survival with increased number of blood transfusions. New Engl. J. Med. *299:* 799–803 (1978).

215 Opelz, G.; Mickey, M., and Terasaki, P.I.: HL-A and kidney transplants: reexamination. Transplantation *17:* 371–382 (1974).

216 O'Reilly, R.J.; Dupont, B.; Pahwa, S.; Grimes, E.; Smithwick, E.M.; Pahwa, R.; Schwartz, S.; Hansen, J.A.; Siegal, R.P.; Sorell, M.; Svejgaard, A.; Jersild, C.; Thomsen, M.; Platz, P.; L'Esperance, P., and Good, R.A.: Reconstitution in severe combined immunodeficiency by transplantation of marrow from an unrelated donor. New Engl. J. Med. *297:* 1311–1318 (1977).

217 Oriol, R.; Cartron, J.; Yvart, J.; Bedrossian, J.; Duboust, A.; Bariety, J.; Gluckman, J.C., and Gagnadoux, M.F.: The Lewis system: new histocompatibility antigens in renal transplantation. Lancet *i:* 574–577 (1978).

218 Parham, P.; Alpert, B.N.; Orr, H.T., and Strominger, J.L.: Carbohydrate moiety of HLA antigens. J. biol. Chem. 252: 7555–7567 (1977).

219 Parham, P. and Bodmer, W.F.: Monoclonal antibody to a human histocompatibility alloantigen, HLA-A2. Nature, Lond. 276: 397–399 (1978).

220 Payne, R.; Tripp, M.; Wigle, J.; Bodmer, W., and Bodmer, J.: A new leukocyte isoantigen system in man. Cold Spring Harb. Symp. quant. Biol. 29: 285–295 (1964).

221 Payne, R. and Rolfs, M.R.: Fetomaternal leukocyte incompatibility. J. clin. Invest. 37: 1756–1763 (1958).

222 Perkins, H.A.; Gantan, Z.; Siegel, S.; Howell, E.; Belzer, F.O., and Kountz, S.L.: Reactions of kidney cells with cytotoxic antisera. Possible evidence for kidney-specific antigens. Tissue Antigens 5: 88–98 (1975).

223 Persijn, G.G.; Gabb, B.W.; Leeuwen, A. van; Nagtegall, A.; Hoogeboom, J., and Rood, J.J. van: Matching for HLA antigens of A, B, and DR loci in renal transplantation by Eurotransplant. Lancet i: 1278–1281 (1978).

224 Peterson, P.A.; Rask, L.; Sege, K.; Klareskog, L.; Anundi, H., and Östberg, L.: Evolutionary relationship between immunoglobulins and transplantation antigens. Proc. natn. Acad. Sci. USA 72: 1612–1616 (1975).

225 Pfizenmaier, K.; Trostman, H.; Röllinghoff, M., and Wagner, H.: Temporary presence of self-reactive cytotoxic T lymphocytes during murine lymphocytic choriomeningitis. Nature, Lond. 258: 238–240 (1975).

226 Piazza, A.: Haplotypes and linkage disequilibrium from the three-locus phenotypes. Histocompatibility testing 1975, pp. 923–927 (Munksgaard, Copenhagen).

227 Piazza, A. and Galfré, G.: A new statistical approach for MLC typing: a clustering technique. Histocompatibility testing 1975, pp. 552–556 (Munksgaard, Copenhagen).

228 Piazza, A.; Sgaramella-Zonta, L.; Gluckman, P., and Cavalli-Sforza, L.L.: The fifth histocompatibility workshop gene frequency data. A phylogenetic analysis. Tissue Antigens 5: 445–463 (1975).

229 Platz, P.; Dupont, B.; Fog, T.; Ryder, L.P.; Thomsen, M.; Svejgaard, A., and Jersild, C.: MLC determinants, measles infection and multiple sclerosis. Proc. R. Soc. Med. 67: 1133–1136 (1974).

230 Platz, P.; Jersild, C.; Thomsen, M.; Svejgaard, A.; Fog, T.; Midholm, S.; Raun, N.; Hansen, S.K., and Grob, P.: Transfer factor treatment of patients with multiple sclerosis. II. Immunological parameters in a long-term clinical trial; in Transfer factor. Basic properties and clinical applications, pp. 649–662 (Academic Press, New York 1976).

231 Race, R.R. and Sanger, R.: Blood groups in man; 6th ed. (Blackwell, Oxford 1975).

232 Rapaport, F.T.: The biological significance of cross-reactions between histocompatibility antigens and antigens of bacterial and/or heterologous mammalian origin; in Transplantation antigens, pp. 181–208 (Academic Press, New York 1972).

233 Redman, C.W.G.; Bodmer, J.G.; Bodmer, W.F.; Beilin, L.J., and Bonnar, J.: HLA antigens in severe pre-eclampsia. Lancet ii: 397–399 (1978).

234 Reisfeld, R.A.; Pellegrino, M.A., and Kahan, B.D.: Salt extraction of soluble HLA antigens. Science, N.Y. 172: 1134–1136 (1971).

235 Reports from a mixed lymphocyte culture workshop. Tissue Antigens 4: 453–525 (1974).

236 Richiardi, P.; Castagnetot, M.; D'Amaro, J.; Schreuder, I.; Vassalli, P., and Curtoni, E.S.: Four new HL-A allelic factors subtypic to HL-A12 and W15. Their correlation with w4 and w6. J. Immunogenet. 1: 323–335 (1974).

237 Rittner, C.H.; Hauptmann, G.; Grosshans, E., and Mayr, S.: Linkage between HL-A (major histocompatibility complex) and genes controlling the synthesis of the fourth

component of complement. Histocompatibility testing 1975, pp. 945–954 (Munksgaard, Copenhagen).

238 Rogentine, C.N. and Plocinik, B.A.: Application of the ^{51}Cr cytotoxicity technique to the analysis of human lymphocyte isoantigens. Transplantation 5: 1323–1333 (1967).

239 Roger, J.H.; Vreeswijk, W. van; D'Amaro, J., and Balner, H.: The major histocompatibility complex of Rhesus monkeys. IX. Current concepts of serology and genetics of Ia antigens. Tissue Antigens 11: 163–180 (1978).

240 Roitt, I.: Essential immunology; 2nd ed. (Blackwell, Oxford 1974).

241 Rood, J.J. van: Leucocyte grouping. A method and its application (Pasmans, Haag 1962).

242 Rood, J.J. van: A proposal for international cooperation in organ transplantation. Histocompatibility testing 1967, pp. 451–452 (Munksgaard, Copenhagen).

243 Rood, J.J. van; Hooff, J.P. van, and Keuning, J.J.: Disease predisposition, immune responsiveness and the fine structure of the HL-A supergene. Transplantn Rev. 22: 75–104 (1975).

244 Rood, J.J. van; Leeuwen, A. van; Bruning, J.W., and Eernisse, J.G.: Current status of human leukocyte groups. Ann. N.Y. Acad. Sci. 129: 446–466 (1966).

245 Rood, J.J. van; Leeuwen, A. van, and Eernisse, J.G.: Leucocyte antibodies in sera from pregnant women. Nature, Lond. 181: 1735–1736 (1958).

246 Rood, J.J. van; Leeuwen, A. van; Keuning, J.J., and Blussé van Oud Alblas, A.: The serological recognition of the human MLC determinants using a modified cytotoxicity technique. Tissue Antigens 5: 73–79 (1975).

247 Rood, J.J. van; Leeuwen, A. van; Schippers, A.; Vooys, W.H.; Frederiks, H.; Balner, H., and Eernisse, J.G.: Leukocyte groups, the normal lymphocyte transfer test and homograft sensitivity. Histocompatibility testing 1965, pp. 37–50 (Munksgaard, Copenhagen).

248 Rosenthal, A.S.: Determinant selection and macrophage function in genetic control of the immune response. Immunol. Rev. 40: 136–152 (1978).

249 Rosenthal, A.S. and Shevach, E.M.: The function of macrophages in antigen recognition by guinea pig T lymphocytes. I. Requirement for histocompatible macrophages and lymphocytes. J. exp. Med. 138: 1194 (1973).

250 Rubinstein, P.; Suciu-Foca, N., and Nicholson, J.F.: Genetics of juvenile diabetes mellitus. A recessive gene closely linked to HLA-D and with 50 percent penetrance. New Engl. J. Med. 297: 1036–1040 (1977).

251 Russell, T.J.; Schultes, L.M., and Kuban, D.J.: Histocompatibility (HL-A) antigens associated with psoriasis. New Engl. J. Med. 287: 738–739 (1972).

252 Ryder, L.P.; Andersen, E., and Svejgaard, A.: An HLA map of Europe. Hum. Hered. 28: 171–200 (1978).

253 Ryder, L.P.; Andersen, E., and Svejgaard, A.: HLA and disease. 3rd Report from the HLA and Disease Registry. Tissue Antigens (in press).

254 Ryder, L.P.; Nielsen, L. Staub, and Svejgaard, A.: Associations between HL-A histocompatibility antigens and non-malignant diseases. Humangenetik 25: 251–264 (1974).

255 Ryder, L.P. and Svejgaard, A.: Histocompatibility associated diseases; in Loor and Roelants, T and B lymphocytes in immune responses, pp. 437–456 (Wiley, London 1977).

256 Ryder, L.P.; Thomsen, M.; Platz, P., and Svejgaard, A.: Data reduction in LD-typing. Histocompatibility testing 1975, pp. 557–562 (Munksgaard, Copenhagen).

257 Sachs, J.A.; Festenstein, H.; Tuffnell, V.A., and Paris, A.M.J.: Collaborative scheme for tissue typing and matching in renal transplantation. IX. Effect of HLA-A, -B, and -D

locus matching, pretransplant transfusion, and other factors on 612 cadaver renal grafts. Transplantn Proc. *9:* 483–486 (1977).

258 Sandberg, L.; Thorsby, E.; Kissmeyer-Nielsen, F., and Lindholm, A.: Evidence of a third sublocus within the HL-A chromosomal region. Histocompatibility testing 1970, pp. 165–169 (Munksgaard, Copenhagen).

259 Sanderson, A. and Batchelor, R.: Lymphocytotoxic reactions of human isoantisera detected by the release of chromium-51 label or by dye exclusion. Histocompatibility testing 1967, pp. 367–369 (Munksgaard, Copenhagen).

260 Sasportes, M.; Fredelizi, D.; Nunez-Roldan, A.; Wollman, E.; Giannogpoulos, Z., and Dausset, J.: Analysis of stimulating products involved in primary and secondary allogenic proliferation in man. I. Preponderant role of the Ia-like DR (Ly-Li) antigens as stimulating products in secondary allogenic response in man. Immunogenetics *6:* 29–42 (1978).

261 Scandiatransplant. HL-A matching and kidney-graft survival. Lancet *i:* 240–242 (1975).

262 Schlosstein, L.; Terasaki, P.I.; Bluestone, R., and Pearson, C.M.: High association of an HL-A antigen, W27, with ankylosing spondylitis. New Engl. J. Med. *288:* 704–706 (1973).

263 Schreiner, G.F. and Unanue, E.R.: Capping and the lymphocyte: models for membrane reorganization. J. Immun. *119:* 1549–1551 (1977).

264 Shearer, G.M.: Cell-mediated cytotoxicity to trinitrophenyl-modified syngeneic lymphocytes. Eur. J. Immunol. *4:* 527 (1974).

265 Sheehy, M.J.; Sondel, P.M.; Bach, M.L.; Wank, R., and Bach, F.H.: HLA LD (lymphocyte defined) typing: a rapid assay with primed lymphocytes. Science, N.Y. *188:* 1308–1310 (1975).

266 Shevach, E.M. and Rosenthal, A.S.: The function of macrophages in antigen recognition of guinea pig T lymphocytes. II. Role of the macrophage in the regulation of genetic control of the immune response. J. exp. Med. *138:* 1213 (1973).

267 Shreffler, D.C. and David, C.S.: The H-2 major histocompatibility complex and the I immune response region. Genetic variation, function, and organization. Adv. Immunol. *20:* 125–195 (1975).

268 Shulman, N.R.; Aster, R.H.; Pearson, H.A., and Miller, H.C.: Immunoreactions involving platelets. VI. Reactions of maternal isoantibodies responsible for neonatal purpura. Differentiation of a second platelet antigen system. J. clin. Invest. *41:* 1059–1069 (1962).

269 Shulman, N.R.; Marder, V.J.; Hiller, M.C., and Collier, E.M.: Platelet and leukocyte isoantigens and their antibodies: serologic, physiologic, and clinical studies. Prog. Haemat. *4:* 222–304 (1964).

270 Simonsen, M.: Artificial production of immunological tolerance. Induced tolerance to heterologous cells and induced susceptibility to virus. Nature, Lond. *175:* 763 (1955).

271 Simpson, E. and Gordon, R.D.: Responsiveness to HY antigen Ir gene complementation and target cell specificity. Immunol. Rev. *35:* 59–75 (1977).

272 Singal, D.P. and Blajchman, M.A.: Histocompatibility (HL-A) antigens, lymphocytotoxic antibodies and tissue antibodies in patients with diabetes mellitus. Diabetes *22:* 429–432 (1973).

273 Smith, M.; Gold, P.; Freedman, S.O., and Shuster, J.: Studies of the linkage relationship of β_2-microglobulin in man-mouse somatic cell hybrids. Ann. hum. Genet. *39:* 21–31 (1975).

274 Snary, D.; Goodfellow, P.; Bodmer, W.F., and Crumpton, M.J.: Evidence against a

dimeric structure for membrane-bound HLA antigens. Nature, Lond. *258:* 240–242 (1975).

275 Snell, G.D.: The H-2 locus of the mouse. Observations and speculations concerning its comparative genetics and its polymorphism. Folia biol. *14:* 335–358 (1968).

276 Snell, G.D.: T cells, T cell recognition structures, and the major histocompatibility complex. Immunol. Rev. *38:* 3–69 (1978).

277 Snell, G.D.; Dausset, J., and Nathenson, S.: Histocompatibility (Academic Press, New York 1976).

278 Solheim, B.G.; Bratlie, A.; Sandberg, L.; Nielsen, L. Staub, and Thorsby, E.: Further evidence of a third HL-A locus. Tissue Antigens *3:* 439–463 (1973).

279 Solheim, B.G.; Engebretsen, T.E.; Flatmark, A.; Jervell, J.; Enger, E., and Thorsby, E.: The influence of HLA-A, -B, -C and -D matching of kidney graft survival. Scand. J. Urol. Nephrol., suppl. 42, pp. 28–31 (1977).

280 Someren, H. van; Westerveld, A.; Hagemeijer, A.; Mees, J.R.; Khan, P.M., and Zaalberg, O.B.: Human antigen and enzyme markers in man chinese hamster somatic cell hybrids. Evidence for synteny between the HL-A, PGM_3, ME_1, and IPO-B loci. Proc. natn. Acad. Sci. USA *71:* 962–965 (1974).

281 Sondel, P.M.; Sheehy, M.J.; Bach, M.L., and Bach, F.H.: The secondary stimulation test (SST). A rapid LD matching technique. Histocompatibility testing 1975, pp. 581–583 (Munksgaard, Copenhagen).

282 Soulier, J.-P.; Moullec, J. et Prou-Wartelle, O.: Recherches de paternité par le système HL-A. Revue fr. Transf. *15:* 11–35 (1972).

283 Springer, T.A. and Strominger, J.L.: Detergent-soluble HLA antigens contain a hydrophilic region at the COOH-terminus and a penultimate hydrophobic region. Proc. natn. Acad. Sci. USA *73:* 2481–2485 (1976).

284 Stocker, J.W.; Garotta, G.; Hausmann, B.; Trucco, M., and Ceppellini, R.: Separation of human cells bearing HLA-DR antigens using a monoclonal antibody rosetting method. Tissue Antigens (in press).

285 Storb, R.; Prentice, R.L., and Thomas, E.D.: Marrow transplantation for treatment of aplastic anemia. An analysis of factors associated with graft rejection. New Engl. J. Med. *296:* 61–66 (1977).

286 Storb, R.; Thomas, E.D.; Buckner, C.D.; Clift, R.A.; Johnson, F.L.; Fefer, A.; Glucksberg, H.; Giblett, E.R.; Lerner, K.G., and Neiman, P.: Allogeneic marrow grafting for treatment of aplastic anemia. Blood *43:* 157–162 (1974).

287 Svejgaard, A.: Iso-antigenic systems on human blood platelets – a survey. Ser. haematol. *2* (1969).

288 Svejgaard, A.: Synergistic action of HL-A isoantibodies. Nature, Lond. *222:* 94–95 (1969).

289 Svejgaard, A.; Bratlie, A.; Hedin, P.J.; Högman, C.; Jersild, C.; Kissmeyer-Nielsen, F.; Lindblom, B.; Lindholm, A.; Löw, B.; Messeter, L.; Möller, E.; Sandberg, L.; Nielsen, L. Staub, and Thorsby, E.: The recombination fraction of the HL-A system. Tissue Antigens *1:* 81–88 (1971).

290 Svejgaard, A.; Christy, M.; Nerup, J.; Platz, P.; Ryder, L.P., and Thomsen, M.: HLA and autoimmune disease with special reference to the genetics of insulin-dependent diabetes; in Rose, Bigazzi and Warner, Genetic control of autoimmune disease, pp. 101–112 (Elsevier North-Holland, New York 1978).

291 Svejgaard, A.; Hauge, M.; Kissmeyer-Nielsen, F., and Thorsby, E.: HL-A haplotype frequencies in Denmark and Norway. Tissue Antigens *1:* 184–195 (1971).

292 Svejgaard, A.; Jersild, C.; Nielsen, L. Staub, and Bodmer, W.F.: HL-A antigens and disease. Statistical and genetical considerations. Tissue Antigens *4:* 95–105 (1974).

293 Svejgaard, A. and Kissmeyer-Nielsen, F.: Cross-reactive human HL-A isoantibodies. Nature, Lond. *219:* 868–869 (1968).

294 Svejgaard, A.; Nielsen, L. Staub; Ryder, L.P.; Kissmeyer-Nielsen, F.; Sandberg, L.; Lindholm, A., and Thorsby, E.: Subdivisions of HL-A antigens. Evidence of a 'new' segregant series. Histocompatibility testing 1972, pp. 465–473 (Munksgaard, Copenhagen).

295 Svejgaard, A.; Platz, P.; Ryder, L.P.; Nielsen, L. Staub, and Thomsen, M.: HL-A and disease associations – a survey. Transplantn Rev. *22:* 3–43 (1975).

296 Svejgaard, A. and Ryder, L.P.: Interaction of HLA molecules with non-immunological ligands as an explanation of HLA and disease associations. Lancet *ii:* 547–549 (1976).

297 Svejgaard, A. and Ryder, L.P.: HLA markers and disease; in Sing and Skolnick, Genetic analysis of common diseases: Applications to predictive factors in coronary heart disease (Alan Liss Publ., in press).

298 Svejgaard, A.; Thorsby, E.; Hauge, M., and Kissmeyer-Nielsen, F.: Genetics of the HL-A system. A family study. Vox Sang. *18:* 97–133 (1970).

299 Sørensen, S.F.: The mixed lymphocyte culture interaction – techniques and immunogenetics. Acta path. microbiol. scand., suppl. 230, pp. 1–82 (1972).

300 Sørensen, S.F. and Nielsen, L. Staub: The genetic basis for reactivity in human mixed lymphocyte culture. Acta path. microbiol. scand. *78:* 719–725 (1970).

301 Tada, T.; Taniguchi, M., and David, C.S.: Properties of the antigen-specific suppressive T-cell factor in the regulation of antibody response of the mouse. IV. Special subregion assignment of the gene(s) that codes for the suppressive T-cell factor in the H-2 histocompatibility complex. J. exp. Med. *144:* 713–725 (1976).

302 Teisberg, P.; Olaisen, B.; Jonassen, R.; Gedde-Dahl, T., and Thorsby, E.: The genetic polymorphism of the fourth component of human complement: methodological aspects and a presentation of linkage and association data relevant to its localization in the HLA region. J. exp. Med. *146:* 1380–1389 (1977).

303 Terasaki, P.I. and McClelland, J.D.: Microdroplet assay of human serum cytotoxins. Nature, Lond. *204:* 998–1000 (1964).

304 Thomas, E.D.; Bryant, J.I.; Buckner, C.D.; Clift, R.A.; Fefer, A.; Johnson, F.L.; Neiman, P.E.; Ramberg, R.E., and Storb, R.: Leukemic transformation of engrafted human marrow cells *in vivo.* Lancet *i:* 1310–1313 (1972).

305 Thomas, E.D.; Buckner, C.D.; Rudolph, R.H.; Fefer, A.; Storb, R.; Neiman, P.E.; Bryant, J.I.; Chard, R.L.; Clift, R.A.; Epstein, R.B.; Fialkow, P.J.; Funk, D.D.; Giblett, E.R.; Lerner, K.G.; Reynolds, F.A., and Slichter, S.: Allogeneic marrow grafting for hematologic malignancy using HL-A matched donor-recipient sibling pairs. Blood *38:* 267–287 (1971).

306 Thomsen, M.; Dickmeiss, E.; Jakobsen, B.K.; Platz, P.; Ryder, L.P., and Svejgaard, A.: Low responsiveness in MLC induced by certain HLA-A antigens on the stimulator cells. Tissue Antigens *11:* 449–456 (1978).

307 Thomsen, M.; Hansen, G.S.; Svejgaard, A.; Jersild, C.; Hansen, J.A.; Good, R.A., and Dupont, B.: Mixed lymphocyte culture technique. Tissue Antigens *4:* 493–506 (1974).

308 Thomsen, M.; Hansen, H.E., and Dickmeiss, E.: MLC and CML studies in the family of a pair of HLA haploidentical chimeric twins. Scand. J. Immunol. *6:* 523–528 (1977).

309 Thomsen, M.; Jakobsen, B.; Platz, P.; Ryder, L.P.; Nielsen, L. Staub, and Svejgaard, A.: LD-typing, polymorphism of MLC determinants. Histocompatibility testing 1975, pp. 509–518 (Munksgaard, Copenhagen).

310 Thomsen, M.; Morling, N.; Ladefoged, J.; Løkkegård, H.; Svejgaard, A., and Thaysen, J.H.: HLA-D and MLC studies in cadaver kidney transplantation (in preparation).

311 Thomsen, M.; Morling, N.; Platz, P.; Ryder, L.P.; Nielsen, L. Staub, and Svejgaard, A.: Specific lack of responsiveness to certain HLA-D (MLC) determinants with notes on primed lymphocyte typing (PLT). Transplantn Proc. *8:* 455–459 (1976).

312 Thomsen, M.; Platz, P.; Andersen, O. Ortved; Christy, M.; Lyngsøe, J.; Nerup, J.; Rasmussen, K.; Ryder, L.P.; Nielsen, L. Staub, and Svejgaard, A.: MLC typing in juvenile diabetes mellitus and idiopathic Addison's disease. Transplantn Rev. *22:* 125–147 (1975).

313 Thomson, G. and Bodmer, W.: The genetic analysis of HLA and disease associations; in Dausset and Svejgaard, HLA and disease, pp. 84–93 (Munksgaard, Copenhagen 1977).

314 Thorsby, E.: The human major histocompatibility system. Transplantn Rev. *18:* 51–129 (1974).

315 Thorsby, E.; Bondevik, H.; Helgesen, A., and Hirschberg, H.: Lymphocyte activating determinants of the human major histocompatibility system (MHS). Transplantn Proc. *7:* 87–92 (1975).

316 Thorsby, E.; Hirschberg, H., and Helgesen, A.: A second locus determining human MLC response. Separate lymphocyte populations recognize the products of each different MLC-locus allele in allogeneic combination. Transplantn Proc. *5:* 1523–1528 (1973).

317 Thorsby, E.; Kissmeyer-Nielsen, F., and Svejgaard, A.: New alleles of the HL-A system. Serological and genetic studies. Histocompatibility testing 1970, pp. 137–151 (Munksgaard, Copenhagen).

318 Thorsby, E.; Segaard, E.; Solem, J.H., and Kornstad, L.: The frequency of major histocompatibility antigens (SD & LD) in thyreotoxicosis. Tissue Antigens *6:* 54–55 (1975).

319 Ting, A. and Morris, P.J.: Matching for B-cell antigens of the HLA-DR series in cadaver renal transplantation. Lancet *i:* 575–577 (1978).

320 Ting, A. and Terasaki, P.I.: Lymphocyte-dependent antibody cross-matching for transplant patients. Lancet *i:* 304–306 (1975).

321 Tosi, R.M.; Pellegrino, M.; Scudeller, G., and Ceppellini, R.: The agar fluorochromasia cytotoxicity test for cell typing. Histocompatibility testing 1967, pp. 351–356 (Munksgaard, Copenhagen).

322 Transplantation Reviews: Surface antigens on nucleated cells, vol. 6 (Munksgaard, Copenhagen 1971).

323 Transplantation Reviews: Suppressor T lymphocytes, vol. 26 (Munksgaard, Copenhagen 1975).

324 Transplantation Reviews: β_2-Microglobulin and HL-A antigen, vol. 21 (Munksgaard, Copenhagen 1974).

325 Transplantation Reviews: HL-A and disease, vol. 22 (Munksgaard, Copenhagen 1975).

326 Trinchieri, G.; Marchi, M. de; Mayr, W.; Savi, M., and Ceppellini, R.: Lymphocyte antibody lymphocytolytic interaction (LALI) with special emphasis on HL-A. Transplantn Proc. *5:* 1631–1646 (1973).

327 Troup, G.M.; Jameson, J.; Thomsen, M.; Svejgaard, A., and Walford, R.L.: Studies of HLA alloantigens of the Navaja Indians of North America. I. Variance of association between HLA-DRW (WIA) and HLA-DW specificities. Tissue Antigens *12:* 44–57 (1978).

328 Tweel, J.G. van den; Blussé van Oud; Alblas, A.; Keuning, J.J.; Goulmy, E.; Termijtelen, A.; Bach, M.L., and Rood, J.J. van: Typing for MLC (LD). I. Lymphocytes from cousin marriages offspring as typing cells. Transplantn Proc. *5:* 1535–1538 (1973).

329 Villano, B.C. del; Nave, B.; Croker, B.P.; Lerner, R.A., and Dixon, F.J.: The oncorna-

virus glycoprotein gp69/71: a constituent of the surface of normal and malignant thymocytes. J. exp. Med. *141:* 172–187 (1975).

330 Vladutiu, A.O. and Rose, N.R.: Autoimmune murine thyroiditis. Relation to histocompatibility (H-2) type. Science, N.Y. *174:* 1137–1139 (1971).

331 Vogel, F. und Helmbold, W.: Blutgruppen-Populationsgenetik und Statistik; in Humangenetik, vol. 1/4, pp. 129–571 (Thieme, Stuttgart 1972).

332 Walford, R.L.; Gossett, T.; Smith, G.S.; Zeller, E., and Wilkinson, J.: A new alloantigenic system on human lymphocytes. Tissue Antigens *5:* 196–204 (1975).

333 Walford, R.L.; Smith, G.S., and Waters, H.: Histocompatibility systems and disease states with particular reference to cancer. Transplantn Rev. *7:* 78–111 (1971).

334 Weitkamp, L.R. and Guttormsen, S.A.: Genetic linkage of a locus for erythrocyte glyoxalase (GLO) with HLA and Bf. Human Gene Mapping 3. Baltimore Conference (1975). Birth Defects Orig. Article Ser., vol. XII, No. 7, pp. 364–366 (1976).

335 WHO: Nomenclature for factors of the HL-A system. Bull. Wld Hlth Org. *39:* 483–486 (1968).

336 WHO: Nomenclature for factors of the HLA system. Histocompatibility testing 1975, pp. 5–20 (Munksgaard, Copenhagen).

337 WHO: Nomenclature for factors of the HLA system. Bull. Wld Hlth Org. *54:* 461–465 (1978).

338 Wiman, K.; Curman, B.; Forsum, U.; Klareskog, L.; Malmnäs-Tjernlund, U.; Rask, L.; Trägårdh, L., and Peterson, P.A.: Occurrence of Ia antigens on tissues of non-lymphoid origin. Nature, Lond. *276:* 711–713 (1978).

339 Woolf, B.: On estimating the relation between blood group and disease. Ann. hum. Genet. *19:* 251–253 (1955).

340 Wunderlich, J.R.; Rogentine, G.N., and Yankee, R.A.: Rapid *in vitro* detection of cellular immunity in man against freshly explanted allogeneic cells. Transplantation *13:* 31–37 (1972).

341 Yamashita, U. and Shevach, E.M.: The histocompatibility restrictions on macrophage T-helper cell interaction determine the histocompatibility restrictions on T-helper cell B-cell interaction. J. exp. Med. *148:* 1171–1185 (1978).

342 Yamazaki, K.; Boyse, E.A.; Miké, V.; Thaler, H.T.; Mathieson, B.J.; Abbott, J.; Boyse, J.; Zayas, Z.A., and Thomas, L.: Control of mating preferences in mice by genes in the major histocompatibility complex. J. exp. Med. *144:* 1324 (1976).

343 Yankee, R.A.; Graff, K.S.; Dowling, R., and Henderson, E.S.: Selection of unrelated compatible platelet donors by lymphocyte HL-A matching. New Engl. J. Med. *288:* 760–764 (1973).

344 Yunis, E.J.; Plate, M.; Ward, F.E.; Seigler, H.F., and Amos, D.B.: Anomalous MLR responsiveness among siblings. Transplantn Proc. *3:* 118–126 (1971).

345 Ziegler, J.B.; Alper, C.A., and Balner, H.: Properdin factor B and histocompatibility loci linked in the rhesus monkey. Nature, Lond. *254:* 609–611 (1975).

346 Zinkernagel, R.M.; Callahan, G.N.; Klein, J., and Dennert, G.: Cytotoxic T cells learn specificity for self H-2 during differentiation in the thymus. Nature, Lond. *271:* 251–253 (1978).

347 Zinkernagel, R.M. and Doherty, P.C.: Restriction of *in vitro* T cell-mediated cytotoxicity in lymphocytic choriomeningitis within a syngeneic or semiallogeneic system. Nature, Lond. *248:* 701–702 (1974).

13. Subject Index